A PLUME BOOK

INNOVATE LIKE EDISON

"People who haven't considered Thomas Edison since high sch
in a new light, so to speak, after reading *Innovate Like Edison*."

—*USA Today*

"*Innovate Like Edison* fully captures the inspiration—perspiration—and best practices required for innovation success today. This book offers you a step-by-step blueprint on how to incorporate this critical thinking into your life and business."
—Dr. Peter H. Diamandis, chairman and founder, X PRIZE Foundation

"In the current global competitive environment, innovation is the key to individual and business success. This book is a must-read for business and innovation leaders for their personal and business success."
—Surinder Kumar, Ph.D., MBA; senior vice president and chief innovation officer, Wm. Wrigley Jr. Company; coauthor of *Riding the Blue Train*

"Leaders, managers, and workers in any industry will find here a story and process for not only learning about business innovation, but practical guidance for implementing it. If you're involved in business or government in any way, you should read this book."
—James G. Clawson, professor, Darden Graduate School of Business Administration, University of Virginia

"Highly readable and full of practical wisdom that you can put to use immediately. Much as you may think you know about Edison and the process of innovation, this book will surprise and delight you."
—Dr. Rajendra S. Sisodia, professor of marketing, Bentley College; coauthor of *Firms of Endearment* and *The Rule of Three*

"By drawing on the wealth of documents available in the Edison archive, Michael J. Gelb and Sarah Miller Caldicott have succeeded admirably in showing us how Thomas Edison operated as an ingenious inventor and a sophisticated director of research and development . . . Accessible to anyone looking for practical advice."
—Paul Israel, director and general editor of the Thomas A. Edison Papers at Rutgers University; author of *Edison: A Life of Invention*

"Reading *Innovate Like Edison* from start to finish on a flight from coast to coast, I landed with three pages of actions to take that will bring innovation to every corner of our business."

—Cal Wick, CEO and founder of Fort Hill Company; coauthor of
The Six Disciplines of Breakthrough Learning

"An engaging, accessible, and fascinating look at one of history's most inventive inventors."

—Leonard Shlain, author of *Art & Physics*, *The Alphabet Versus the Goddess*,
and *Sex, Time, and Power*

"This is an urgently important book. The only competitive advantage, whatever our field of endeavor, will be our ability to innovate, and this is the concrete, 'how to' 25-step blueprint for putting all of us, including our nation, back on this critical path. I'm not just reading, but studying this book with my children!"

—Verne Harnish, founder, Entrepreneurs' Organization (EO), and CEO,
Gazelles Inc.; author of *Mastering the Rockefeller Habits*

"Precisely what high tech needs today. Gelb and Caldicott have articulated the core fiber of innovation culture for companies to naturally innovate and compete . . . Required reading for executives, managers, and engineers alike."

—Andrew J. Stein, chief marketing officer, Paradigm, BV

"This powerful and insightful book will inspire you with the brilliance of what Edison achieved. Much more than that, it will tell you how he actually did it. This book shines a bright and searching light on his daily habits of innovation. It will open your eyes."

—Sir Ken Robinson, author of *Out of Our Minds: Learning to Be Creative*

ALSO BY MICHAEL J. GELB

Body Learning:
An Introduction to the Alexander Technique

Present Yourself!

Lessons From the Art of Juggling: How to Achieve Your Full Potential
in Business, Learning, and Life (with Tony Buzan)

Thinking for a Change:
Discovering the Power to Create, Communicate and Lead

Samurai Chess:
Mastering Strategic Thinking Through the Martial Art of the Mind
(with Raymond Keene)

How to Think Like Leonardo da Vinci:
Seven Steps to Genius Every Day

The How to Think Like Leonardo da Vinci Workbook:
Your Personal Companion to How to Think Like Leonardo da Vinci

Discover Your Genius:
How to Think Like History's Ten Most Revolutionary Minds

More Balls Than Hands:
Juggling Your Way to Success by Learning to Love Your Mistakes

Da Vinci Decoded:
Discovering the Spiritual Secrets of Leonardo's Seven Principles

INNOVATE LIKE EDISON

*The Five-Step System for
Breakthrough Business Success*

Michael J. Gelb
and Sarah Miller Caldicott

A PLUME BOOK

PLUME
Published by the Penguin Group

Penguin Group (USA) Inc., 375 Hudson Street, New York, New York 10014, U.S.A. • Penguin Group (Canada), 90 Eglinton Avenue East, Suite 700, Toronto, Ontario, Canada M4P 2Y3 (a division of Pearson Penguin Canada Inc.) • Penguin Books Ltd., 80 Strand, London WC2R 0RL, England • Penguin Ireland, 25 St. Stephen's Green, Dublin 2, Ireland (a division of Penguin Books Ltd.) • Penguin Group (Australia), 250 Camberwell Road, Camberwell, Victoria 3124, Australia (a division of Pearson Australia Group Pty. Ltd.) • Penguin Books India Pvt. Ltd., 11 Community Centre, Panchsheel Park, New Delhi – 110 017, India • Penguin Group (NZ), 67 Apollo Drive, Rosedale, North Shore 0632, New Zealand (a division of Pearson New Zealand Ltd.) • Penguin Books (South Africa) (Pty.) Ltd., 24 Sturdee Avenue, Rosebank, Johannesburg 2196, South Africa

Penguin Books Ltd., Registered Offices: 80 Strand, London WC2R 0RL, England

Published by Plume, a member of Penguin Group (USA) Inc. Previously published in a Dutton edition.

First Plume Printing, November 2008
10 9 8 7 6 5 4 3

The Library of Congress has catalogued the Dutton edition as follows:
Gelb, Michael.
    Innovate like Edison : the success system of America's greatest inventor / Michael J. Gelb and Sarah Miller Caldicott.
        p. cm.
    ISBN 978-0-525-95031-8 (hc.)
    ISBN 978-0-452-28982-6 (pbk.)
    1. Creative ability in technology. 2. Technological innovations. 3. Edison, Thomas A. (Thomas Alva), 1847–1931. I. Caldicott, Sarah Miller. II. Title.
    T49.5.G45 2007
    608—dc22   2007026653

Printed in the United States of America
Original hardcover design by Joseph Rutt

PUBLISHER'S NOTE
While the authors have made every effort to provide accurate telephone numbers and Internet addresses at the time of publication, neither the publisher nor the authors assumes any responsibility for errors, or for changes that occur after publication. Further, publisher does not have any control over and does not assume any responsibility for author or third-party websites or their content.

BOOKS ARE AVAILABLE AT QUANTITY DISCOUNTS WHEN USED TO PROMOTE PRODUCTS OR SERVICES. FOR INFORMATION PLEASE WRITE TO PREMIUM MARKETING DIVISION, PENGUIN GROUP (USA) INC., 375 HUDSON STREET, NEW YORK, NEW YORK 10014.

From Michael:
To the light and love of my life, Deborah Domanski.

From Sarah:
With deepest love to my children,
Nicholas and Connor,
and
To my great-great-grandfather
Lewis Miller and his enterprising daughter,
Mina Miller Edison.

From Michael and Sarah:
To the fulfillment of Edison's vision
of human progress through the
application of his timeless approach
to innovation.

CONTENTS

ACKNOWLEDGMENTS

This book emerged as a natural expression of the principles that it describes. We are passionate about our goal to bring Edison's practical wisdom to you, and this was vividly clear from our first conversation. The competencies and elements for *Innovate Like Edison* that we describe in the following pages guided us through our entire creative process. We delighted especially in the experience of master-mind collaboration with each other, and with an extraordinary network of individuals who contributed to this book.

Our special thanks for the support we received from Jim D'Agostino, Dr. Annaliesa Anderson, Ed Bassett, Michael Boehm, Tony Buzan, Debbie Happy Cohen, Leslie "Duck" Copland, John Fogler, Joan and Sandy Gelb, Desiree Gruber, Yolanda Harris, Mary Hogan, Dr. Marvin Hyett, Mitch Hoffman, Joel Jaffe, Grandmaster Raymond Keene, O.B.E., Dr. Ken Koblan, Dr. Surinder Kumar, Debra Kurtz, the entire Miller family, Jeff Monda, Vanda North, Steve Odland, Dr. Dennis Perman, Tom Quick, the entire Reading family, Susan RoAne, Wendy Rothman, Dr. Dale Schusterman, Dr. Richard Sheridan, Robert Tangora, Michael Thieneman, Dr. Win Wenger, Dr. John Wai, Diana Whitney, and Michael Wing.

Our very special thanks to the individuals and organizations who offered research assistance and ongoing inspiration, including Professor James Clawson of the Darden School of Business at the University of Virginia; Jeanie Egmon, Director, Complexity In Action Network at Northwestern University and Professor of Managerial Economics and Decision Sciences, Kellogg School of Management; Kimball C. Firestone; Professor Vijay Govindarajan and Associate Professor Christopher Trimble at the Amos Tuck School of Business at Dartmouth; Leonard de Graaf of the Edison National Historic Site; Pamela Miner at the Edison and Ford Winter Estates in Fort Myers, Florida; Nancy Munro; Rini Paiva at the National Inventors Hall of Fame; Jon Schmitz at the Chautauqua Institution in Chautauqua, New York; The Strategic Forum—New York; The Strategic Forum—South

xi

Florida; Karen Strouse; Bob Troyer; team leaders Carla Withrow and Josh Loyer, and distinguished student Derek Bartie, and all the graduate students who served on the Edison Research Support Team at the H. Wayne Huizenga School of Business and Entrepreneurship, including Lisette Bennett, Erika Burney, Patricia Campbell, Steve Cooke, Lisa Fox, Syreeta Joseph, Efthimia Karipedes, Sunita Kaul, Judi Li, Jeannette Porras, Franklin Ramchandani, and Kelly Ticlavilca.

Our gratitude to the excellent team at Dutton, especially Stephen Morrow, Erika Imranyi, Lenny Telesca, Melanie Gold, and Trena Keating. And to Muriel Nellis and Jane Roberts of the Literary and Creative Artists Agency.

Our deepest thanks to those individuals who made exceptional contributions to the development of this book, and who stood as supreme examples of the Edisonian spirit: Dr. Paul Israel, Director and General Editor, The Thomas A. Edison Papers at Rutgers University; Dr. Curtis Carlson, President and CEO of SRI International; National Inventors Hall of Fame inductees Dr. Helen M. Free, Dr. Robert E. Kahn, Dr. Donald B. Keck, Dr. Robert S. Langer, Jr., Dr. James E. West, and Nancy Miller Arnn.

As part of our research we interviewed many innovators from a wide range of disciplines. Stories, examples, and quotations from these remarkable individuals appear throughout the text. In most cases we've cited the name of each source and the name of their organization in the Resources and Reference Notes (page 263). In a few instances, the sources wished to remain anonymous. In those cases, we've included their quotations and stories without mentioning names.

In true Edisonian fashion, we developed enough material in writing one book to fill three. As a result, we have included additional resources, quotes, and examples of Edison's innovations on our Web site: *www.innovatelikeedison.com*. We welcome you to extend your interest in Edison by visiting the site.

We strongly recommend the definitive biography by Paul Israel entitled *Edison: A Life of Invention*, which represents the most comprehensive biographical review of Edison's life written to date.

We also recommend The Project Gutenberg eBook: *Edison, His Life and Inventions* by Frank Lewis Dyer and Thomas Commerford Martin, contemporaneous biographers of Thomas Edison.

FOREWORD

Thomas Edison is often described as the ultimate inventor. After all, he invented the electric light bulb, the phonograph, moving pictures, and much more. Many also imagine that he was a "lone genius" toiling away in a sterile laboratory with nothing to offer the world other than his remarkable inventions. What could we possibly learn from a person like this, who worked more than a hundred years ago?

Innovate Like Edison proves that the clichéd vision of Edison is wrong. It demonstrates that we can all learn a tremendous amount from him. Gelb and Caldicott show that Edison is actually a stimulating role model for innovation, entrepreneurship, and personal success, not just invention. And, as we enter the highly competitive global economy, his example is even more relevant today.

There have been many great inventors, but it was only Edison who managed to tackle one new marketplace after another with world-changing products and services. And this much is certain: You do not achieve this level of sustained success by accident. Edison was working in a much more productive manner compared to his competitors. This book describes Edison's approach to innovation in a simple, direct way that makes it clear how all of us can benefit from his ideas.

Innovation is the creation and delivery of new customer value in the marketplace. It is not just invention, creativity, or teamwork. Until a new product or service is in the hands of a customer, it is just another potentially clever idea, not an innovation. And to be sustainable, it must provide sufficient profit to the enterprise producing it. Edison was an innovator. Yes, he invented. But that was only part of his process for creating compelling new value for his customers.

Edison's passion for creating successful innovations led him to study and invent processes to increase his chances of success—what we now call "innovation best practices." For example, Edison was the first to assemble in a single enterprise the resources needed to develop not just one, but a host of new products and services.

By doing so he invented the modern research laboratory, which is one of the most important innovations of all time.

Edison had a comprehensive approach to innovation. He aspired to address major market needs. He formed multidisciplinary teams to develop his new products. He created manufacturing companies to produce these products. And he developed new business models to monetize them. These are not the traits of "just" an inventor. These are the traits of a disciplined innovator. Consequently, Edison's innovation best practices apply to all of us, whether we are a CEO, a recent college graduate, or someone spending our free time attempting to turn a new idea into a product.

Innovate Like Edison outlines the history of Edison's life, his great achievements, and the people who inspired and competed with him. Like all great people, Edison also made big mistakes. Many times he violated his own innovation best practices. He infamously refused to accept the advantages of alternating current versus direct current, and he stubbornly held on to the phonographic cylinder after the superior disc was introduced. These events just emphasize how important it is, even if you are an Edison, to stay true to the fundamentals of innovation.

Innovate Like Edison traces Edison's progress in developing and using innovation best practices. For example, before he entered a new field he meticulously did his homework. He read everything that was available. He also made sure he could visualize a practical path to success. After all, the light bulb was just one part of a complete electrical distribution system. Marketplace success required the ability to economically develop generators, electrical plugs, insulation, power meters, and hundreds of other parts. To help organize Edison's practices, Gelb and Caldicott introduce the concept of "innovation literacy." It is a way to test the extent of your innovation skills while providing a framework to add new ones.

In 2006 my colleague William W. Wilmot and I published the book *Innovation: The Five Disciplines for Creating What Customers Want*. The five disciplines we describe are based on leading major organizations in the process of innovation and in studying exemplars of innovation, like Edison.

One of our goals in writing the book was to make clear that innovation is a discipline to be studied and improved. Just as important, the book includes fundamental ingredients necessary for success. Curiously, many people have the mistaken belief that the words "discipline" and "innovation" are contradictions. They can be, of course. But, as Edison demonstrated, when you have the right skills and embrace essential practices, you actually liberate your creativity and greatly improve your ability to innovate. Without these skills and practices, your chance of success drops precipitously. Gelb and Caldicott have crafted a practical guide that shows you how to bring some of these innovation best practices to your organization.

One of our colleagues, Douglas Engelbart, the SRI International inventor of the computer mouse, windows, hypertext, and many other essential attributes of the personal computer, has cleverly said, "The better we get, the better we get at getting better." That was Edison. He was driven to get better. You can get better too. *Innovate Like Edison* will help show you the way.

Curtis R. Carlson, Ph.D.
President and CEO
SRI International
Menlo Park, California

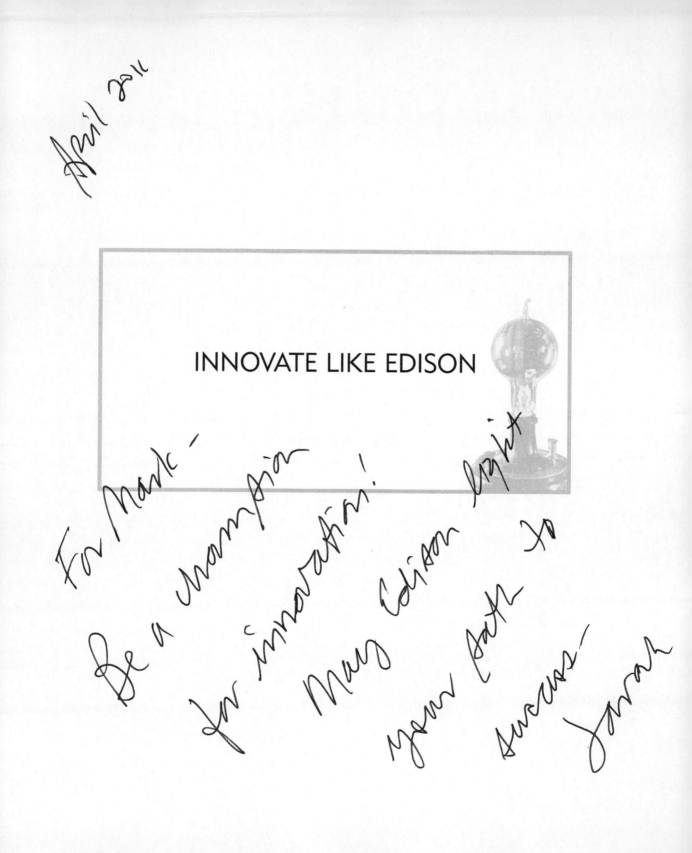

April 2016

INNOVATE LIKE EDISON

For Mark —
Be a champion
for innovation!
May Edison light
your path to
success —
Sarah

My philosophy of life is work—bringing out the secrets of nature and applying them for the happiness of man. I know of no better service to render during the short time we are in this world.
    —*Thomas Alva Edison*

PART ONE

AN AMERICAN INNOVATOR
WHO CHANGED THE WORLD

Chapter One

INTRODUCTION: TURNING ON THE LIGHT

> If we all did the things we are capable of doing we would literally astound ourselves.
>   —*Thomas Edison*

At one thirty A.M. on October 22, 1879, everything was ready for the astounding experiment that would change the world forever. Thomas Edison, age thirty-two, and his colleagues Charles Batchelor and Francis Jehl, huddled around a series of glass tubes, gauges, and wires suspended on a tall wooden stand in the middle of Edison's Menlo Park, New Jersey, laboratory. Edison focused intently on a prototype, pear-shaped light bulb mounted at the upper edge of the stand, carefully checking the partial vacuum seal at its base. Protruding from the hand-blown glass bulb were two razor-thin platinum lead-in wires, connected inside the bulb to a carbonized cotton filament no thicker than a human hair. Eyeing the lamp for a moment, Edison was satisfied that the connections holding the bulb's fragile filament in place were intact. Edison asked Jehl, "Are you ready?"

"Ready," he replied.

Jehl began evacuating oxygen from the glass lamp by pouring mercury into a long tube at the top of the stand. As mercury flowed through the tube, it slowly drew oxygen out of the lamp, creating a vacuum. They watched as large oxygen bubbles percolated through the viscous liquid, gradually giving way to smaller bubbles as the inner atmosphere of the lamp was voided of air. To hasten creation of the vacuum, Edison lit an alcohol burner, drawing the small, steady flame across the exterior of the glass bulb, warming it to remove any moisture within the bulb. Large bubbles suddenly surged into the mercury as heat from the burner drove more air out of the bulb.

Activating a battery sitting on a nearby table, Edison connected a wire running from one pole of the battery to one of the razor-thin platinum lead-in wires, taking a second wire from the battery's other pole and briefly touching it against the second lead-in wire. Current flowed upward, warming the filament, its red glow gradually ridding the carbonized cotton of any gases that could alter the vacuum. Edison repeated this filament-warming process several times until no more bubbles were visible in the mercury.

Pleased, Edison called upon his glass blower, Ludwig Boehm, to fully seal off the bulb at its base, preserving the near-perfect vacuum they had just created—one part per million—a breakthrough achieved by the Menlo Park staff just a few months earlier.

Once the bulb had been fully sealed, Edison placed it in a small stand on the table. He bound one wire running from each pole of the battery to each of the free platinum lead-in wires from the bulb, creating a complete circuit. Current surged upward through the high-resistance, carbonized cotton thread. The room filled with light.

Over the next several hours, Edison, Batchelor, Jehl, and Boehm were joined at intervals by other members of Edison's inner circle and support staff: John Kruesi, the master prototype builder who could construct anything from even the sketchiest of Edison's drawings; Francis Upton, the well-traveled mathematician and physicist who formalized Edison's visionary concepts into equations; plus John Lawson and Martin Force, laboratory assistants. All were present over the course of that historic day.

After six hours, Jehl whispered to his compatriots, "It's still burning!" This filament had now outlasted all previous efforts. Consulting his pocket watch for confirmation, Francis Upton smiled, while Batchelor and Edison continued fashioning additional carbonized filaments in case unexpected problems arose.

At three P.M.—13.5 hours into the experiment—Edison turned up the battery's voltage, making the filament glow even more brightly. At four P.M. Edison and his team watched as the glass lamp cracked, and the last flickers of the filament's incandescent brilliance were spent. The light had shone for 14.5 hours, including a full one-hour stress test. A jubilant Edison declared, "If it will burn that number of hours now, I know I can make it burn a hundred!" The practical incandescent electric light had been born.

Edison's breakthrough was the product of not just one, but five inventions: the development of an improved vacuum process; a thin, high-resistance filament

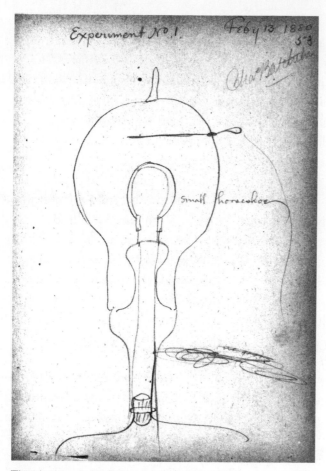

This drawing from Edison's notebook, dated February 13, 1880, and witnessed by Charles Batchelor, illustrates a new filament designed to drive longer burn-life.

This drawing of the incandescent electric light was registered by Edison with the U.S. Patent Office in November 1879.

made of cotton thread, carbonized to sustain burn-life; platinum lead-in wires that could bring electric current to the filament; a method for holding the filament in place; connecting all of these technologies together in a near-perfect vacuum environment, inside a hand-blown glass light bulb. Edison's Menlo Park team would push the technological boundaries of their discovery even further less than one year later, when they began commercially manufacturing the first incandescent light bulb at the Edison Lamp Works. Less than two years later, on September 4,

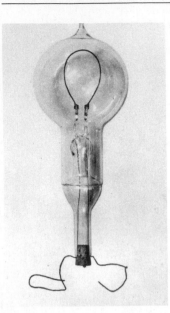

This photograph of a vacuum-sealed test light bulb reveals the internal intricacies of the filament, the delicate clamps holding the filament in place, as well as the exterior platinum wires to which a battery could be connected for testing.

1882, the world's first central power system would illuminate New York City from Edison's Pearl Street station, near Wall Street. The age of electrical power had begun.

If you're reading this book with the help of an electric light, it's worth pausing to remember that it was Edison's remarkable innovation that makes it possible. Edison's landmark success with the incandescent light bulb and his development of an entirely new system for distributing and monitoring the application of electrical power changed the world forever in a profoundly practical and profitable way. It is fitting that the light bulb is now a universal metaphor for "bright ideas," "brilliance," and "ingenious thinking." It is also fitting that Edison is the extraordinary genius behind that icon.

Edison didn't just invent the incandescent light, he envisioned and created a system for lighting the entire world. His system included the means for funding, producing, distributing, monitoring, marketing, and continuously improving his inventions. Then, he did the same thing with the phonograph and moving pictures, launching the modern entertainment industry.

When asked what his rules were for the laboratory and its staff, Edison loved to respond, "Hell, there are no rules here—we're trying to accomplish something." However, as we study his life and work, it becomes clear that he *did* have powerful rules for innovation. These rules were manifest in his establishment of the world's first Research and Development laboratory at Menlo Park, New Jersey, and the first Industrial Research and Development (R&D) complex at West Orange, New Jersey. Edison was the first person to create a system for innovation. His focus on practical accomplishment set the stage for America's global leadership in innovation. Before Thomas Edison, innovation was viewed as the random product of a lone genius. Edison was, of course, an exceptional genius, but the greatest product of his genius was the establishment of a systematic approach to success that he believed anyone could emulate.

Principles of personal success and business innovation went hand in hand throughout Edison's life. Napoleon Hill (1883–1970), author of the self-improvement classic *Think and Grow Rich*, based his work on interviews with many of the most prominent people of his time, including Edison. In *Innovate Like Edison*, we revisit Thomas Edison, and find a timeless inspiration beneath the clichés that history has built over him. We will guide you to apply the essential elements of Edison's approach to personal success and business innovation now.

THE INNOVATION IMPERATIVE

> Every organization—not just business—needs one core competence: innovation.
> —*Peter Drucker*

Thomas Edison was the most outstanding figure in an era marked by an extraordinary confluence of American innovation—including the work of Alexander Graham Bell, Henry Ford, George Eastman, Harvey Firestone, John D. Rockefeller, George Westinghouse and Andrew Carnegie—that accelerated America's leadership in global business. In the last ten years, however, growing challenges from China, India, South Korea, and other emerging economies are placing pressure on American preeminence.

Many recent commentators have cautioned that America's longtime dominance of key technologies and product design processes is waning. According to noted author Thomas Friedman, much of our U.S.-centric thinking about key computer hardware, software, cellular, and medical technologies is outdated. Friedman argues that businesses everywhere have now entered a fundamentally new era of competition where only the innovative survive.

Recognizing the breadth and magnitude of this challenge, the U.S. Council on Competitiveness, a group comprised of business executives, academics, and leaders drawn from a wide range of American life, recently released their National Innovation Survey. The survey makes it clear that global innovation leadership has begun shifting away from the United States.

The council's report notes that as of 2005:

- The United States is now home to only six of the world's top twenty-five most innovative information technology companies. This represents a decline of more than 100 percent over the last thirty years.

- In 2003, the United States fell behind China as the number one recipient of global business investment.

- Sweden, Finland, Israel, Japan, and South Korea all invest a greater percentage of GDP in R&D than the United States.

- Federal funding of basic research is expected to decline in real terms through 2010.

How did this happen? Although the American R&D powerhouse initiated by Edison has changed the face of the world with decades of inventions and innovations, few fundamental changes have been made to the structure and process of most R&D departments since World War II. Reed Business Information's *R&D Magazine* indicated in a January 2005 article that R&D staffing hierarchies remain virtually unchanged since the mid-1950s. And unchanged staffing hierarchies often reflect a less than creative attitude toward this "innovation imperative." Moreover, innovation is no longer just the province of R&D, it must be lived and breathed by everyone, in every function, at every level in the organization.

How can America reinvent its powerhouse innovation machine to respond to increasing global competition?

How can your organization succeed in its quest to translate this critical strategic priority into practical implementation?

Most important, how can you develop the abilities you need to be a leader in this quest?

We believe that the starting point for answering all these questions can be drawn from our own heritage. Thomas Edison invented systematic innovation and there is much to be gained by revisiting his methods and making them relevant to the challenges we face now.

Of course, as Edison understood, innovation is much more than invention. Through the establishment of his two extraordinary laboratories

On the fiftieth anniversary of Edison's invention of the incandescent electric light, he was presented with this commemorative replica by Henry Ford at Greenfield Village in Dearborn, Michigan, where Ford had fully reconstructed Edison's Menlo Park laboratory.

at Menlo Park and West Orange, New Jersey, Edison drove innovation on many levels, including strategic, technological, product/service, process, and design innovations.

How did Edison excel in so many different kinds of innovation? What are the key elements of his thought process and how can we apply them to our most important innovation challenges? How did Edison develop his remarkably resilient, creative, and optimistic attitude to life, and how can we apply his approach to our personal success? How did he find the right people to hire, and why did he choose the collaborators he did? What techniques did Edison use to test his ideas, commercialize them, scale them up, and market them? What are the implicit "rules" of Edison's approach, the underlying system for innovation and success that drove his unique, unprecedented, and still-unmatched productive genius?

And how can you apply his methods in your life and your organization now?

In the pages that follow we will answer these questions as we guide you to *Innovate Like Edison.*

INNOVATION LITERACY

Imagine you are an English teacher, and your principal asks you and your students to put on a play by Shakespeare. Your job will, of course, be much easier if your students can read and write. If your students are illiterate—if they can't read their parts and can't write down their notes for developing their roles—the play will be very difficult to produce. If you are a leader in an organization engaging in an innovation initiative—perhaps implementing strategy developed with the help of consultants—and your people are innovation illiterates; i.e., if they are unfamiliar with the practical attitudes, thinking, and communication skills necessary for innovation, then innovation is unlikely to result.

Literacy is traditionally understood as the ability to read and write. Recently, however, the definition has grown to embrace broader concepts of communication fluency in a context of individual empowerment. The National Institute for Literacy now defines it as "an individual's ability to read, write, speak in English, compute and solve problems at levels of proficiency necessary to function on the job, in the family of the individual and in society." The United Nations Educational, Scientific and Cultural Organization (UNESCO) offers an even more expansive definition: "Literacy is the ability to identify, understand, interpret, create, communicate and compute, using printed and written materials associated with varying contexts. Literacy involves a continuum of learning to enable an individual to achieve his or her goals, to develop his or her knowledge and potential, and to participate fully in the wider society."

We believe that this "continuum of learning to enable an individual to achieve his or her goals" must now include a working knowledge of the attitudes, strategies, and skills of innovation. We call this working knowledge *innovation literacy*. We aren't alone. We asked Vijay Govindarajan, professor of international business at the Amos Tuck School of Business at Dartmouth, "What's the single greatest point of leverage for individuals and organizations who want to succeed in a competitive business environment?" He responded:

> We've got to get every member of the organization, from top to bottom, literate in innovation just like we make them literate in finance, or literate in marketing, or literate in any other management disciplines. Innovation is not only about ideas and creativity, it's a whole discipline about how you turn an idea into reality. Innovation literacy has to be across the board. It's got to be done.

If you are functioning as a leader now or if you aspire to be a leader in the future you must become innovation literate. The good news is that innovation literacy is something you can learn. As Edison emphasized, we are all gifted with astounding potential. In the following pages we will teach you Edison's system for translating that potential into achievement.

EDISON'S FIVE COMPETENCIES OF INNOVATION

Innovate Like Edison presents Edison's essential approach to innovation success. His approach is based on what we call the Five Competencies of Innovation. The five competencies are comprised of a total of twenty-five elements—or building blocks—that support them. The five competencies and twenty-five elements represent a core curriculum for you to achieve innovation literacy. The competencies will enable you to *Innovate Like Edison*, whatever your current level of innovation experience. If you are expected, as part of your job, to be innovative everyday, we are confident you will find that Edison's inspiration and approach will complement and enliven your endeavor. If you come home and unleash your most creative self after regular work hours, this book is also for you. If you are new to innovation, then we believe there's no better way to get started. *Innovate Like Edison* is a guidebook enabling you to thrive in a world that increasingly rewards innovation.

Edison was the original master of the rigorous, disciplined innovation process, but he also was a wizard when it came to promoting and maintaining a culture of innovation. The cultural side of innovation often provides the missing link to real implementation. For Whirlpool, a company competing in a historically low-growth industry with numerous international players, a renewed focus on innovation has vaulted the firm onto the Top 100 list of the world's most innovative companies as ranked by *Business Week* in 2006. As Mike Thieneman, executive vice president and chief technology officer at Whirlpool, stated in a recent conversation with the authors: "We've been putting all the tools and processes of innovation we need into place over the past five years, but we don't have 'the soft side' of innovation down yet." The "soft side" is all about the personal attitudes, communication skills, ways of thinking, and the management culture that supports these.

For Edison, innovation existed both at the level of the individual and as a force of science, learning, strategy, and culture that engages our humanity and our entrepreneurial spirit. *Innovate Like Edison* is therefore designed to guide you in how to *do* innovation, and how to *be* it. This book is also inspired by the Edisonian notion that the principles of personal success and organizational innovation are inseparable. Personal success and fulfillment requires you to learn how to think like an innovator; and, for your organization to be successful, innovation is now more important than ever.

The five competencies of innovation constitute the heart of this book. The term "competency" as used here refers to bundles of skills that must be mastered to achieve success in a particular area of endeavor. It draws on the notion of "core competence" introduced by leading strategist Gary Hamel.

Edison's five competencies of innovation are:

1. Solution-centered Mindset
2. Kaleidoscopic Thinking
3. Full-spectrum Engagement
4. Master-mind Collaboration
5. Super-value Creation

#1—SOLUTION-CENTERED MINDSET

The phenomenon of seeing what we expect or want to see is called "mindset." It functions all the time, consciously or unconsciously, for better or worse. For example, if you decide you want to buy a hybrid vehicle and ask yourself, "What's the best hybrid for me?" you can be sure that the next time you're out on the road or walking through a parking lot you will notice the Toyota Prius, Saturn VUE, Ford Escape, and Lexus GS hybrids. You will also notice articles and advertisements in newspapers and magazines related to energy-efficient vehicles that you would have probably overlooked before you set your mind on "hybrid."

Your mindset reflects your sense of purpose, and your sense of purpose organizes your perceptions. In other words, purpose determines perception. As psychologist Abraham Maslow observed, "People who are only good with a hammer see every problem as a nail." A solution-centered mindset gives you access to a wide range of tools for innovating.

Edison's purpose was clear: "bringing out the secrets of nature and applying them for the happiness of man." He believed that his success was inevitable and this belief energized his every endeavor. Edison's unwavering focus on finding solutions allowed him to embrace incredibly complex challenges and overcome many setbacks. His *solution-centered mindset* allowed him to embrace seemingly fantastic goals, like lighting the world, and make them come true. Edison aligned his goals with his passions and cultivated a powerful sense of optimism that had a magnetic, positive effect on his coworkers, investors, customers, and ultimately the entire nation. We call it *charismatic optimism*.

Edison's passion for his goals and his charis-

This page, in Edison's handwriting, was taken from one of more than two hundred notebooks Edison and Francis Upton generated to compare the price of electricity and gas.

matic optimism were nurtured by an unrelenting desire to learn, especially by read-ing. Throughout his life, Edison devoured books, plays, journals, magazines, scientific papers, and newspapers. Edison's voracious reading created a constant stream of ideas, insights, and inspiration that led him to breakthrough solutions. His never-ending quest for greater depth and breadth of knowledge helped him develop an unprecedented approach to experimentation in service of innovation. His experi-ments were characterized by a remarkable combination of persistence and rigorous objectivity that accelerated his success.

A *solution-centered mindset* is the launching pad for the realization of your most ambitious innovation objectives and the fulfillment of your highest personal aspi-rations. We will guide you to discover how Edison developed this competency, and how you can, too.

#2—KALEIDOSCOPIC THINKING

Edison's ability to generate a vast range of ideas drove his world-beating approach to practical solution finding. He could consider many problems at the same time and was able to look at each one from multiple angles. At the height of his explora-tion into electrical power, for example, he worked on forty projects simultane-ously. Edison credited his remarkable facility for making creative connections to his "mental kaleidoscope."

Kaleidoscopic thinking is our term for his unparalleled approach to practical creativity. You will learn Edison's strategies for juggling multiple projects and how to "turn a problem around" from every angle. You'll develop your ability to gener-ate ideas, make creative connections, and discern patterns. Using both your imag-ination and your reasoning ability, you will discover how to liberate your mind from the constraints of habitual thinking. And, like Isaac Newton, Albert Ein-stein, and Leonardo da Vinci, Edison cultivated the use of metaphors, analogies, and visual thinking. His down-to-earth way of picturing things in his mind's eye and then on paper is surprisingly easy to learn and it will become an invaluable part of your innovation toolkit.

#3—FULL-SPECTRUM ENGAGEMENT

When you are overworked and stressed out it's very difficult to focus effectively on innovation. How can you successfully manage a massive workload, like Edison did,

Dec 21, 1879 NY Herald

EDISON'S LIGHT.

The Great Inventor's Triumph in Electric Illumination.

A SCRAP OF PAPER.

It Makes a Light, Without Gas or Flame, Cheaper Than Oil.

TRANSFORMED IN THE FURNACE.

Complete Details of the Perfected Carbon Lamp.

FIFTEEN MONTHS OF TOIL.

Story of His Tireless Experiments with Lamps, Burners and Generators.

SUCCESS IN A COTTON THREAD.

The Wizard's Byplay, with Bodily Pain and Gold "Tailings."

HISTORY OF ELECTRIC LIGHTING.

The near approach of the first public exhibition of Edison's long looked for electric light, announced to take place on New Year's Eve at Menlo Park, on which occasion that place will be illuminated with the new light, has revived public interest in the great inventor's work, and throughout the civilized world scientists and people generally are anxiously awaiting the result. From the beginning of his experiments in electric lighting to the present time Mr. Edison has kept his laboratory guardedly closed, and no authoritative account (except that published in the HERALD some months ago relating to his first patent) of any of the important steps of his progress has been made public—a course of procedure the inventor found absolutely necessary for his own protection. The HERALD is now, however, enabled to present to its readers a full and accurate account of his work from its inception to its completion.

Headline from a full-page December 21, 1879, *New York Herald* article.

without succumbing to exhaustion and burnout? Time management isn't the answer. Edison understood that although time on the clock was limited, the wellspring of creative inspiration was boundless. He drew on a seemingly endless source of energy, and he had a remarkable range of expression. No matter what he was doing, he was fully engaged, living life in the present. His ability to move freely, efficiently, passionately, and creatively through a day's many activities and roles was a critical aspect of his success method. Edison discovered an optimal rhythm to facilitate amazing stamina and high performance. We call his approach *full-spectrum engagement*. It is a competency that you can cultivate to access the same boundless energy that fueled Edison. You will learn his approach to balancing work and play, solitude and collaboration, concentration and relaxation.

#4—MASTER-MIND COLLABORATION

Napoleon Hill introduced the concept of master-mind groups nearly a century ago. For Hill, the master-mind group was more than just the product of the synergy that results from effective teamwork. He defined the "master mind" as: "coordination of knowledge and effort in a spirit of harmony, between two or more people, for the attainment of a definite purpose." Hill believed that when people come together under the right circumstances, they can multiply their individual brainpower in an expanding framework of positive, creative energy. Hill's ideas about the nature of mind were inspired by Thomas Edison's speculations on the nature of thought, matter, and energy. And the functioning of a higher "group intelligence" was clearly evident in Edison's approach to building innovation teams capable of *master-mind collaboration*. You'll learn how to create and leverage that "higher group intelligence."

#5—SUPER-VALUE CREATION

Edison learned early in his career that creativity and invention were necessary, but not sufficient for innovation. He discovered that the most rewarding kinds of creativity came from ideas targeted to markets. Edison understood that creativity had to be translated into commerce and that successful commerce was based on meeting the needs of his customers. Creating value for others is the core of personal happiness and success. Helping others is life's most fulfilling endeavor. It's great to follow your bliss and do what you love, but if you truly want the money to follow, then you must strategically link your bliss to something that others, preferably those with discretionary income, want or need.

Before the advent of the modern disciplines we call "marketing" and "branding," Edison developed a systematic approach to understanding, measuring, and fulfilling the needs of his customers. He developed original strategies for quantifying and intuiting the potential value his customers would find in his products and services. Thus, he was able to consistently trump his competitors by providing significantly greater value to his customers. *Super-value creation* is our term for this Edison competency. We will show you how he did it and help you apply his approach to your innovation process now.

The five competencies are based on what Edison actually did and how he thought about the process of innovation. We will set the stage for your exploration of the competencies with a biographical chapter that maps the influences and inspirations that drove Edison's unique passion for innovation.

This hand-drawn sketch by Francis Upton conveys the optimism and enthusiasm of Edison's Menlo Park team.

In Part 2 we delve into the five competencies. Each competency consists of five building blocks, or *elements*. Each element concludes with a section entitled Creating Innovation Literacy, designed to help you apply what you've just learned about Edison's methods to your own life, starting now. The Creating Innovation Literacy sections contain practical exercises, suggestions for reflection, stories, anecdotes, advice, and references to resources that represent the most important thing you can know or do to translate your understanding of Edison's approach into action. After you've completed this part of the book you will be ready to put Edison's innovation system to work.

Part 3 is about continuing to improve your innovation process as it grows to incorporate others. Here we introduce the Edison Innovation Literacy Blueprint. This tool allows you to quantify the level of innovation literacy you possess, within each competency and overall. It illustrates strengths as well as areas for improvement. A one-page visual schematic of the Edison Innovation Literacy Blueprint is provided to help you integrate everything you will learn and to set goals for your continuing progress. When an organization has a critical mass of individuals who are innovation literate, it begins to form the basis of a viable corporate innovation infrastructure. Corporate innovation infrastructure is strengthened when an organization integrates innovation literacy in its hiring practices, reward systems, outlook on the marketplace, means of testing ideas, and the way it forms teams. The final chapter of *Innovate Like Edison* offers further resources for managing and leading the process of introducing innovation literacy to others in your organization.

Chapter Two

THE STUFF OF DREAMS:
THE LIFE OF THOMAS EDISON
(1847–1931)

> Things may come to those who wait . . . but only the things left by those who hustle.
> —*Abraham Lincoln*
>
> Everything comes to him who hustles while he waits.
> —*Thomas Edison*

EDISON'S WORLD

When Thomas Edison was born, James Knox Polk served as president of the United States and Queen Victoria ruled the British Empire. Edison was fourteen years old when the American Civil War began and he lived for thirteen years after the end of the First World War.

Thomas Edison and Henry Ford championed capitalism in the United States as the Bolsheviks assassinated the czar in Russia. As communism took hold in the Soviet Union, Hitler began his rise to power in Germany in the 1920s. The League of Nations was established in 1919 and the Great Depression began in the United States after the stock market crash of 1929.

The horse was still the fastest means of transportation when Edison was a child growing up in Milan, Ohio, but in the course of his life railways were introduced, then automobiles, submarines, airplanes, helicopters, and rockets. Edison witnessed the introduction of blue jeans (1874), Coca-Cola (1886), jazz (early 1900s), Corn Flakes (1906), the pop-up toaster (1927), and Mickey Mouse (1928). As Edison

generated his world-changing inventions, other innovators brought forth the typewriter (1873), telephone (1876), radio (1901), X-rays (1895), and penicillin (1928).

Charles Darwin (1809–1882), Sigmund Freud (1856–1939), Mohandas Gandhi (1869–1948), Marie Curie (1867–1934), and Albert Einstein (1879–1955) were among Edison's most extraordinary contemporaries.

Of course, when Edison was born, candles, torches, and gas lamps were the only sources of light for homes, public buildings, and streets. If you wanted to hear a speech or a musical performance, you had to be present when it was delivered. And the only "moving pictures" were the ones people conjured in their imaginations. But Edison's imagination changed all that.

He dreamed of a new world, and then he created it.

ARCHETYPE OF INVENTION

"We are such stuff as dreams are made on," Shakespeare wrote in *The Tempest*, and Edison seemed to take it as a commandment. Edison loved Shakespeare's poetry and was fond of quoting him. During lulls in the normally frenetic pace in the telegraphy offices where he worked as a young man, Edison delighted in amusing his colleagues by hunching his back, distorting his gait, and holding forth with the opening of *Richard III*, "Now is the winter of our discontent . . ." Later in his career, when he was dissatisfied with the progress toward the perfection of one of his inventions, he would often jot this same opening line in his notebook.

Although his partial deafness and rather high-pitched voice didn't serve his fantasy of becoming a Shakespearean actor, Edison did comment directly about Shakespeare's influence on his approach to innovation: "My but that man did have ideas! He would have been an inventor, if he had turned his mind to it. He seemed to see the inside of everything. Perfectly wonderful how many things he could think about. His originality in the way of expressing things has never been approached."

It is easy to imagine that if Shakespeare lived in late nineteenth-century America, he would've been inspired to write a play about "The Wizard of Menlo Park." Although Shakespeare wasn't available to write about Edison, a French novelist, August Villiers de L'Isle-Adam, did write a science fiction novel featuring Edison that was published in 1886. Foreshadowing contemporary research into robotics, and reinventing Dr. Frankenstein's monster, L'Isle-Adam has his fictional Edison craft an android named Eve who represents a functional improvement over *Homo sapiens*. As Paul Israel, author of *Edison: A Life of Invention*, comments, "In 1889, the year in which Villiers set his novel, Edison himself was working to reproduce life not by creating it artificially but by capturing living beings through the media of sound recordings and motion pictures."

It's difficult to imagine just how fantastic Edison's innovations were perceived to be by his contemporaries. You can get a feeling for just how amazed people were by Edison's genius through the comment of Daniel Craig, one of the investors in Edison's work, who once wrote to him, "If you should tell me you could make babies by machinery, I shouldn't doubt it."

Edison offers powerful inspiration and practical guidance for anyone interested in innovation. Although he was gifted with exceptional talent as a thinker, inventor, and entrepreneur, he believed that we are all capable of astounding accomplishments if we are willing to work hard. He left clear guidance on how we can leverage our hard work for maximum benefit. We will help you apply his guidance to your own life and work in the next section, but now let's consider Edison's schooling, key aspects of his development, and other elements that help us appreciate the man as well as the legend.

> Peter Drucker called Thomas Edison the "archetype for every high-tech entrepreneur."

AT SCHOOL: LIKE LEONARDO AND EINSTEIN

Like Leonardo da Vinci and Albert Einstein, young Thomas Edison did not fit very well in the standard classroom. If they were youngsters attending school today, all three of these geniuses would probably be diagnosed with attention deficit disorder and given prescriptions for behavior-modifying drugs. The young Leonardo "confounded his schoolmaster" with his unrelenting questioning; Einstein was told by one of his teachers that "he would never amount to anything," and his parents then sent him to a special school; and the young Edison was considered awkward by his teacher, who allegedly described his brain as "addled." Edison's mother, Nancy, insisted on taking him out of the formal classroom, preferring to homeschool her son.

Thomas Edison, age ten.

Fortunately, Nancy Edison was a skilled teacher. She sensed that her son's so-called addled brain and awkward ways were reflections of his extraordinary gifts.

Nancy Edison's confidence in her son proved to be critical in spurring his amazing achievements. Edison attributed the core of much of his success to her belief in him and to the high standards she set. As he commented, "She was always so true and so sure of me . . . And always made me feel I had someone to live for and must not disappoint." Nancy Edison provided more than just encouragement, confidence, and high standards for her son, she gave him a superb education in learning how to learn. As Edison commented in a talk to a group of students in New Jersey in the spring of 1912, "My mother taught me how to read good books quickly and correctly and this opened up a great world in literature. I have always been very thankful for this early training."

In addition to the works of Shakespeare, Edison's reading included *The Decline and Fall of the Roman Empire* by Edward Gibbon and *The History of England* by David Hume. His study of history led him to appreciate the special importance of technological advancement in the development of civilizations. Edison was also strongly influenced by the writings of Thomas Paine. Edison borrowed Paine's *Age of Reason* from his father's bookshelf when he was thirteen and, as he commented many years later, "I can still remember the flash of enlightenment which shone from his pages." It seems that for the young Edison, the enlightenment was the realization that the divine was to be approached and understood through the power of rational inquiry into the natural world, through science. As Paine wrote: "The principles of science lead to this knowledge; for the creator of man is the creator of science, and it is through that medium that man can see God, as it were, face to face."

Edison immersed himself in the science textbooks of his day, and he showed a special interest in chemistry. Although he loved to read, young Thomas liked to experiment even more. He spent every spare penny on lab equipment and chemicals, creating his first chemistry bench in a corner of the basement. A courageous Nancy Edison allowed it, although she feared that Thomas and his friends "would yet blow [their] heads off."

EARLY ENTREPRENEURSHIP

At age twelve, Edison got a job as a newsboy on the Grand Trunk Railway, working the sixty-mile route between Port Huron and Detroit. His entrepreneurial spirit manifested immediately as he employed other boys in businesses selling bread, candy, fruit, vegetables, and other goods along the train's route. He also requested permission to establish an improvised chemical laboratory in the train's baggage compartment, and to fashion a printing press for publishing a small newspaper on the train. Edison called his paper *The Weekly Herald*. He sold it to rail patrons for a few cents per copy.

As Edison worked along the rail route, the Civil War raged. The Battle of Shiloh, on April 6 and 7 of 1862 was a turning point of the conflict and in the young Edison's business life. He conceived the notion of telegraphing the news of such battles and the actions of famous generals and commanders ahead along the train line to raise anticipation and interest for the reports in the paper. And as demand for the paper grew, at each stop he doubled the price from five to ten cents. He quickly realized that "instead of the usual 100 papers I could sell 1000." Edison arranged to borrow money for this venture from the editor of the *Detroit Free Press*. Edison made a big profit and resolved immediately to deepen his mastery of telegraphy and printing. These early experiences as a newspaper publisher and entrepreneur set the stage for Edison's eventual success in promoting his innovations to the world.

Over the last thirty years, thousands of clever, technically oriented young people have gravitated to Silicon Valley and other high-tech centers to seek their fortunes in the emerging field of information technology. When Edison was a young man, the burgeoning field of telegraphy offered the same kind of promise. His acceptance into the fraternity of "lightning jockeys" (also known as "brass pounders" and "lightning slingers") began when he rescued the three-year-old son of telegraph operator James MacKenzie from an onrushing rail car. The grateful MacKenzie asked Edison how he could thank him for his act of bravery, and Edison replied without hesitation, "Telegraphy lessons!"

At age fourteen, Thomas set up a chemistry lab using money he earned selling newspapers on the Grand Trunk Railway.

Edison practiced incessantly and became quite skilled in this highly competitive, new "high-tech" field. Demand for first-rate telegraphers was high and operators moved from place to place seeking adventure and opportunity. For four years (1863–67), Edison honed his skills as a telegrapher and traveled extensively through the Midwest. He made a pilgrimage to the public libraries in all the cities he visited. In addition to staying on top of the technical developments in telegraphy, he sought to learn as much as he could about a wide range of subjects including geology, history, chemistry, and poetry. He devoured newspapers and magazines not just for news of the war, but also to keep abreast of politics, international events, financial trends, and technological advancements. Edison also relished the study of the lives and works of great individuals.

ON THE SHOULDERS OF GIANTS

Great individuals almost always pay tribute to their extraordinary predecessors. Einstein, for example, kept a picture of Sir Isaac Newton above his bed, and Newton said that he would not have seen so far had he not been "standing on the shoulders of giants." Thomas Edison was also inspired by the giants who preceded him. Benjamin Franklin, Michael Faraday, and Abraham Lincoln served as particularly powerful influences.

> I start where the last man left off.
> —*Thomas Edison*

Benjamin Franklin was born in Boston, on January 17, 1706, and died in 1790, fifty-seven years before Edison's birth. Franklin laid the foundations of the American tradition of freedom, self-improvement, and innovation that made Edison's achievements possible. Foreshadowing Edison, Franklin was the embodiment of the words he published in his *Poor Richard's Almanack*, "Energy and persistence conquer all things."

Franklin's discoveries in the field of electricity were revolutionary. Fascinated by electrical experiments he witnessed while visiting his hometown of Boston in 1746, Franklin turned his "energy and persistence" toward an investigation of the subject. Through a series of novel inventions, he amazed his friends by generating tiny flashes of lightning, making a toy spider skitter across the floor, and using electricity to immediately relight candles that had just been blown out.

Franklin began to suspect that lightning and electricity were the same. In June 1752, he conducted his famous kite experiment, charging a key at the end of his kite string with electricity drawn from the lightning above. Recognizing the immense power and danger inherent in this crackling force, Franklin invented the lightning rod as a means to collect electricity for study and to protect buildings, ships, and the people therein. As he experimented and formulated his ideas, Franklin created a new vocabulary for the study and application of electricity. He coined terms that set the stage for Edison's innovations, terms we still use today, such as *battery, conductor, condenser, charge, discharge, uncharged, negative, positive, minus, plus, electric shock,* and *electrician.*

Franklin lived for eighty-four glorious years (same as Edison). When he was twenty-two, he demonstrated his sense of humor and broad philosophical perspective by publishing a prospective epitaph for himself, which read: "The body of B. Franklin, Printer (Like the Cover of an Old Book Its Contents torn Out And Stript of its Lettering and Gilding) Lies Here, Food for Worms. But the Work shall not be Lost; For it Will (as he Believ'd) Appear once More In a New and More Elegant Edition Revised and Corrected By the Author."

One year after Franklin died, the torch of electrical genius was relit with the birth of Michael Faraday. The son of a blacksmith, Faraday was born in London on September 22, 1791 (he died in 1867, when Edison was twenty). At fourteen, he apprenticed as a bookbinder and began to read everything he could about science, and his special passion—chemistry. Inspired, Faraday conducted his own chemical experiments and designed and built his own laboratory. At nineteen, he became a member of the City Philosophical Society, a group devoted to self-improvement through the exploration of scientific matters.

In 1821, Faraday discovered the principle of electromagnetic rotation that ultimately led to the creation of the electric motor. Ten years later, he formalized the principle of electromagnetic induction, which led to the development of the electric transformer and generator. Before Faraday, the scientific community viewed electricity as a bizarre and curious phenomenon. His extraordinary discoveries changed this perspective, and set the stage for a technological revolution. Expanding on the vocabulary introduced by Franklin, Faraday coined the terms *anode, cathode, electrode, electrolyte,* and *ion,* he set the stage for the innovations of Edison and others.

I have got so much to do and life is so short, I am going to hustle.
 —*Edison's comment to a colleague after staying up all night
   reading the works of Faraday*

Faraday prefigured Edison's approach to innovation and also provided his own best epitaph. As he wrote in his notebook: "All this is a dream. Still, examine it by a few experiments. Nothing is too wonderful to be true, if it be consistent with the laws of nature."

"The dogmas of the quiet past are inadequate to the stormy present. The occasion is piled high with difficulty, and we must rise with the occasion. As our case is new, so we must think anew, and act anew." These lines, from Abraham Lincoln's Second Annual Message to Congress, were delivered on December 1, 1862, and serve as an operating statement for the career of Thomas Edison who was fifteen years old at the time. Lincoln provided compelling inspiration for the young entrepreneur. When the teenage Edison created his first newspaper, *The Weekly Herald*, he took every opportunity to fill it with flattering articles about his hero.

Lincoln, the sixteenth president of the United States was born on February 12, 1809, in Kentucky. Lincoln was an inventor and patron of innovation. The first president to receive a transcontinental telegram, Lincoln is also the only U.S. president to apply for and receive a patent for an invention. Lincoln conceived of a portable boat fashioned of leather with inflatable bellows just below the water line, designed for carrying soldiers or munitions across shoals if their vessel ran aground.

Lincoln understood that freedom set the stage for innovation. He wholeheartedly supported the U.S. patent system, which granted its first patent in 1790. Lincoln believed patents created a framework for the nation's progress. The patent system, as he expressed it, "added the fuel of interest to the fire of genius, in the discovery and production of new and

This bronze medallion, designed for the National Inventors Hall of Fame in Akron, Ohio, is bestowed annually upon each inductee, and bears the profiles of both Edison and Lincoln.

useful things." And no one was to combine fuel and fire to create new and useful things like Thomas Edison.

Revered for saving the Union and emancipating the slaves, Lincoln should also be remembered as a trailblazer in America's leadership in innovation. Indeed, the National Inventors Hall of Fame recognizes both Lincoln's and Edison's contributions to American invention. Upon the Hall of Fame's establishment in 1973, Thomas Edison was the first inventor inducted, and three years later, it created a bronze medal for its honorees proudly bearing the profiles of Thomas Edison and Abraham Lincoln.

SELF-INVENTION

Edison is much more than a memory in bronze. He still has an urgent personal message for us. Like Lincoln, Faraday, and Franklin, Edison was "self-made." As much as any other figure in our history, Edison embodies the great American tradition of self-improvement. As Paul Israel commented on the development of Edison's career, "Edison . . . followed the ideology of self-improvement to move through the ranks." The same unstoppable ambition that drove Edison to continuously improve his skill as a telegrapher also inspired him to experiment with ways of improving the technology of his field. He focused on creating better technical tools for training telegraphers and on developing methods of sending multiple messages on one telegraph wire.

In 1868, Edison moved to Boston to begin a job with the Western Union Telegraph Company as a master telegrapher. Working twelve hours a day, six days a week, Edison still found time to work on his own inventions.

As he made his way in the world of communications and telegraph technology, Edison reached a turning point in his career with the inventing and patenting of a vote-counting machine that he attempted to sell to lawmakers in New England. His invention enabled legislators to cast votes from their seats and to have them tallied instantly and accurately. The problem was that the politicians preferred to have a delay between the casting and counting of votes so that lobbying could take place and results might be altered. As one legislator put it, Edison's invention would disturb "the delicate political status quo." His brilliant invention had no market.

Edison turned this disappointment into a lesson that guided all his future endeavors. He pledged that he would "never waste time inventing things that people would not want to buy."

Edison began work on a stock printer, a device that he knew would attract significant commercial interest. He was successful enough in this new enterprise that he was able to resign from his job at Western Union. He placed a notice in the main telegraphy trade journal proclaiming that he would now "devote his time to bringing out his inventions."

Although Boston was a hospitable place for scientific, technical, and cultural discourse, it was relatively conservative when it came to financing the big dreams of a young inventor. In the spring of 1869, Edison moved to New York City. Commenting on the more robust entrepreneurial environment he found in New York, Edison noted, "People here come and buy without your soliciting."

In New York, Edison marketed his services as an inventor-for-hire to major telegraph companies. When his newspaper business had generated profits a decade before, Edison spent the money investing in his own chemistry lab. Now, as money from his invention contracts began to flow, he established a series of experimental machine and manufacturing shops in Newark, New Jersey, to further his inventive work.

After only two years in the New York area, Edison had established a reputation as a leading inventor. He achieved renown as "the best electro-mechanician in the country." During this period, Edison worked on perfecting automatic telegraphy, improving telegraph printers, and the design of fire alarms. He solved the problem of sending multiple messages on one wire, inventing the quadruplex telegraph, which sent two messages simultaneously in each direction. He also invented the electric pen, which Lewis Carroll, author of *Alice's Adventures in Wonderland* (1865), described as "quite the best thing yet invented for taking a number of copies." This remarkable invention led to the development of the Edison Mimeograph Machine, which remained in use until the 1970s, and marked the founding of the A. B. Dick Company, which still exists today. In 1875, Edison's observation of "sparking" in his telegraph instruments led him to formulate a notion he called "the etheric force." Although criticized by journalists, including one who called this an example of "Edison's new moonshine," it turns out that Edison had discovered what would later more formally be known as high-frequency electromagnetic waves.

THE BIRTHPLACE OF SYSTEMATIC INNOVATION

In December of 1875, Thomas Edison purchased two tracts of land in Menlo Park, New Jersey, about thirty-five miles southwest of New York City. On one he established his home, and on the other he established what was to become the first Research and Development laboratory in history. The establishment of Edison's Research and Development laboratory at Menlo Park was a world-changing innovation. Over the next six years, Edison and his teams generated a series of extraordinary inventions at Menlo Park, while creating what Paul Israel terms "a new model for invention that became the cornerstone of modern industrial research."

During this period of amazing productivity, Edison received more than four hundred patents on inventions including the incandescent light bulb and the world's first phonograph. He became famous internationally and was renowned as "The Wizard of Menlo Park." Other innovations generated at Menlo Park included the carbon button transmitter, facilitating a dramatic improvement in the functioning of Alexander Graham Bell's recently invented telephone. The sound quality of Bell's original receiver left listeners straining to hear each caller's words, and Edison's transmitter made the telephone suitable for practical use.

From 1878 to 1881, Edison developed a comprehensive system to support his invention of incandescent electric lighting. During this period of intensive research, he observed a phenomenon later termed "The Edison Effect," which became the basis for vacuum tube electronics. The Edison Effect occurs when a wire is placed between the strands of a filament in a light bulb, serving as a valve to control the flow of electric current. Massive industries based on vacuum tube technology were spawned in the United States during the twentieth century based on applications of the Edison Effect, including radio broadcasting, radar, television, and analog computers.

In 1881, Edison transferred his Menlo Park offices and laboratories to several locations in New York City and the surrounding metropolitan area. Although he was always engaged in many remarkable endeavors simultaneously, his major focus for the next five years was the manufacture, installation, and continuous improvement of his electrical system first in New York—and then around the world.

In 1887, Edison built a much larger Industrial Research laboratory in West Orange, New Jersey, that he intended to be "the best equipped and largest laboratory extant and the facilities superior to any other for rapid and cheap development of an invention." He succeeded. The West Orange laboratory took the concept that Edison originated at Menlo Park to even greater heights, creating

In 1881, at the age of thirty-four, Edison networked with business owners, financiers, and politicians in New York City to gain approval for a live test of the world's first central power station.

what we now call Industrial Research and Development, combining extensive research facilities with major manufacturing operations.

Edison's West Orange facility included laboratories devoted to research in chemistry, physics, and metallurgy and, of course, had an extensive library. The original complex consisted of five buildings with a three-story main lab containing a power plant, machine shops, and testing labs. Edison, always ready to embrace change, expanded and modified the facility to suit his evolving needs and make it possible to work on a remarkably diverse array of projects simultaneously. At its peak during the years of WWI (1914–18) Edison's center spanned more than twenty acres and provided jobs for thousands of people.

At West Orange, Edison and his team of "Muckers" created a complete system to support his invention of the phonograph in the same way that he developed a system to support the workings of the light bulb. With these innovations, he gave birth to the recording industry.

And, then seeking to create something that "does for the eye what the phonograph does for the ear," Edison, with the help of other pioneers, invented the movies. First he developed the technology for silent films and then he experimented with ways to use the phonograph as a soundtrack for silent movies, paving the way for "talking pictures." Edison, with William Kennedy Laurie Dickson, also designed and built the "Black Maria," the world's first movie studio.

"Muckers" is an affectionate term Edison applied to his staff. And he was sometimes referred to as "Chief Mucker." The term "Muckers" is derived from the substance that bound together the ore briquettes Edison manufactured at his mining operations in Ogden, New Jersey, in the early 1890s. It connotes the Edisonian commitment to doing whatever it takes to get the job done.

THE LAUNCH OF GENERAL ELECTRIC AND THE FOUNDATION OF YANKEE STADIUM

In addition to directing his expanding team of muckers in West Orange, Edison initiated businesses around the world to promote, and profit from, his electric lighting innovations. Although he had created a comprehensive system for the practical electrical illumination of the world, the appropriate corporate structure to manage this burgeoning industry proved elusive. In 1889, Edison attempted to bring together the many individual lighting companies he had established. He reorganized these businesses under one umbrella company known as the Edison General Electric Company. However, as competitors arose, and great financiers like J. P. Morgan became involved, Edison's enterprise and those of his main competitors were consolidated into the General Electric Company in 1892. Edison wasn't pleased by the removal of his name from the company title and he expressed his displeasure by exiting the lighting business completely. He remained a major stockholder in the original GE, but sold his shares to generate the capital to invest in a new venture. He proclaimed, "I'm going to do something now so different and so much bigger than anything I've ever done before people will forget that my name ever was connected with anything electrical." Edison's new scheme involved purifying iron ore electromagnetically. Despite years of effort and enormous expense, the Edison Ore Milling Company was a commercial bust. Meanwhile, the shares of GE stock that Edison sold had appreciated to a value of more than $4 million dollars ($82.7 million today). Edison's reaction when he learned of what he might have had? "Well, it's all gone, but we had a hell of a good time spending it."

Fortunately, Edison's work in perfecting the phonograph and developing motion pictures yielded enough profit to keep him afloat financially. As the new century dawned, Edison embraced the challenge of creating a more efficient storage battery for cars and other electrically powered vehicles. He devoted ten years to the development of an alkaline battery, but by the time he made it practical for everyday use, gasoline was just becoming the preferred source for powering automobiles. Nevertheless, Edison's intensive labor was rewarded with significant profits as his battery was adapted for use in powering everything from municipal vehicles to maritime buoys.

Ever resilient, Edison also managed to recoup some of his losses from the ore business by selling his proprietary technologies to other mine owners. And then, he adapted the ore-crushing technology he had devised to make cement. Through his

By 1892, Edison was a world-famous icon.

Edison in work clothes at his New Jersey and Pennsylvania Concentrating Works in 1895.

Edison Portland Cement Company, he also introduced a roasting kiln that became an industry standard. His finely ground cement was used for bridges, dams, and in other major building projects including the construction of Yankee Stadium.

EDISON'S FAMILY LIFE

While Edison was busy inventing the incandescent lighting system, the phonograph, movies, and a new approach to systematic innovation, he also got married—twice—and had six children. Of course, Thomas Edison's first love was his work. Eighteen-hour days were his norm and it wasn't unusual for him to work all night (with a break for "lunch" at midnight). Edison was notorious for forgetting—or just working right through—family anniversaries, birthdays, and social engagements. Like most great geniuses, he wasn't always the most sensitive husband and father. Before considering his relationships with the women who were most significant to him, let's review the basic facts of Edison's family life presented along a simple timeline:

Edison Family Timeline

1847 (February 11) Thomas Alva Edison is born in Milan, Ohio, the youngest of the seven children of Samuel and Nancy Edison

1854 (Spring) The Edison family moves to Port Huron, Michigan

1871 (April 9) Death of Edison's mother, Nancy

1871 (December 25) Edison marries Mary Stillwell

1873 (February 18) Birth of Marion Estelle (nicknamed Dot), Edison's first daughter in Newark, New Jersey

1876 (January 10) Birth of Thomas Alva, Jr. (nicknamed Dash), Edison's first son in Newark, New Jersey

1878 (October 26) Birth of William Leslie, Edison's second son in Menlo Park, New Jersey

1884 (August 9) Death of Mary Stillwell Edison

1886 Edison buys Glenmont, new home in Llewellyn Park, New Jersey

1886 (February 24) Weds Mina Miller in Akron, Ohio

1888 (May 31) Birth of Madeleine, Edison's second daughter

1890 (August 3) Birth of Charles, Edison's third son

1896 (February 26) Death of Edison's father, Samuel, in Norwalk, Ohio

1898 (July 10) Birth of Theodore Miller, Edison's fourth son

1899 (February 17) Death of Lewis Miller, Edison's father-in-law in Akron, Ohio

1900 (March) Begins tradition of annual family vacations in Fort Myers, Florida

1931 (October 18) Edison dies at Glenmont

Edison/Miller Family Tree — Simplified

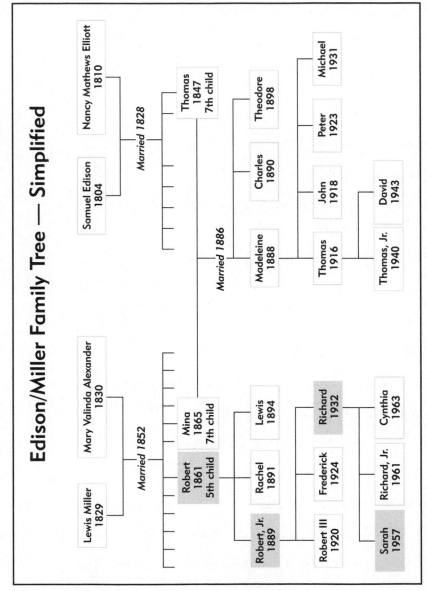

Lewis Miller, inventor of the Buckeye Mower and Reaper, had a delightful rapport with his son-in-law, Thomas Edison. This family tree has been simplified to show the direct descendants of Lewis Miller's son Robert, Mina's older brother and great-grandfather of Sarah Miller Caldicott. More detailed information about the Edison/Miller heritage can be found in the Reference Notes.

EDISON AND WOMEN

Edison's beloved mother, Nancy, died in 1871, when he was twenty-four years old. Later the same year he married Mary Stillwell, an employee at the News Reporting Telegraph Company, a business that Edison co-founded. Their courtship was certainly unconventional by contemporary standards of office behavior. As Edison walked past the workstation of the sixteen-year-old Stillwell, he paused for a moment. She turned around and proclaimed to her boss, "Mr. Edison, I can always tell when you are behind me or near me." Edison replied by asking how she might account for this awareness, and Stillwell answered that she didn't really know. Edison responded by proposing marriage, and eight days later they wed.

It's easy to imagine that Edison was pursuing marriage to fill the void created by the loss of his mother. He and Mary produced three children, but accounts of their relationship suggest that they were a less than ideal match. Edison was disappointed that Mary didn't seem to understand his work, and he complained that she was something of a hypochondriac. In a page of his notebooks, for example, Edison writes his wife's maiden name "Stillwell" a few times before transposing it into "Stillsick." We can also be sure that Mary was disappointed with Edison's long hours and his lack of sympathy for what proved to be her very real health problems. Indeed, Mary Edison died on August 9, 1884. Although he may have been less than attuned to her when they were together, he was devastated by her loss. As his eldest daughter, Marion, later recalled, she discovered her father "shaken with grief, weeping and sobbing so he could hardly tell me that Mother had died in the night."

Thomas Edison was a man's man. He was an alpha male in a thoroughly male-dominated world of engineers, chemists, telegraphers, inventors, scientists, and businessmen. Yet his ideas about the power and potential of women went far beyond the stereotypes of the time. Edison believed that men and women were capable of "equal partnership." He criticized the treatment of women as property and felt that man's "lust for ownership" had created a situation that stifled the full expression of women's intelligence and ability. By forcing women into "petty tasks," men had prevented them from gaining the appropriate "brain exercise" necessary for full development. Always looking to a brighter future, Edison believed that "real sex independence" would allow humanity to evolve to a higher plane. And, he hoped that the new technologics that he was devoted to creating would gradually liberate women from mundane tasks and allow them to develop their brains fully.

Edison's enlightened notions about "equal partnership" and respect for the

Edison in his late fifties.

intelligence and character of women were made manifest in his choice of Mina Miller as his second wife. Mina was the love of his life, and as close to a perfect partner as any great genius has ever had.

Mina was the daughter of Lewis Miller, a brilliant inventor, businessman, and social pioneer. Miller received ninety-two patents for his inventions focusing on the improvement of agricultural equipment, most notably the Buckeye Mower and Reaper. He was inducted into the National Inventors Hall of Fame in May 2006.

Mina met Thomas in New Orleans in January 1885, where she had accompanied her father to the World Industrial and Cotton Centennial Exposition. Lewis was introducing a number of new features for the Buckeye, and Edison's improvements to the telephone were on display in the Bell Telephone exhibit booth. After the exposition, Edison spoke of Mina to his business associate, Ezra Gilliland. A few months later, Gilliland, who lived in Boston, invited Mina—who was attending a women's seminary there—to his home for a social gathering. Edison joined the party, and his friendship with Mina began in earnest.

Lewis Miller was instrumental in founding the famous Chautauqua Institution in Chautauqua, New York, a unique educational community that has continued to thrive for over a century. The Chautauqua movement promotes cultural, philosophical, and ethical education and exchange for adults in a beautiful, recreational environment. Edison wrote of his father-in-law, "He was one of the kindest and most lovable men I ever knew, and spent his life trying to make it possible for all mankind to reach the higher planes of living."

Lewis Miller raised his daughters with the best of liberal education, including tours of Europe and studies in music. In the summer of 1885, Edison traveled to Chautauqua to see Mina, and meet her extended family. Edison was smitten. When they parted, Edison commented to friends that he couldn't stop thinking about "her gentleness and grace of manner and her beauty and strength of mind." Every day without her deepened his yearning. As he noted in his diary, "this celestial mud ball has made another revolution and no photograph yet from the Chautauquain Paragon of Perfection. How much longer will Hope dance on my intellect?"

Romantic pining of this sort is incongruous given Edison's steely focus on invention and business. He was so distracted by the woman he described as "the yardstick for measuring perfection" that upon seeing a woman who resembled her while crossing a street in Boston he "got thinking about Mina and came near being run over by a street car."

In the fall of 1885, Edison invited Mina to accompany him on a trip along the St. Lawrence River in upstate New York, and then to the White Mountains of New Hampshire. It could be said they were on the same wavelength. Their courtship included Thomas teaching Mina Morse code so that they could share intimacies privately while traveling with others. Thomas and Mina tapped out their coded messages on each other's hands. As Edison described it, "We could use pet names without the least embarrassment, although there were three other people in the carriage." The electricity between them continued to develop. Thomas tapped out a special Morse code message to Mina during their carriage ride to New Hampshire: "Will you marry me?" And, to his great delight, she responded immediately, "Yes."

Mina Miller, daughter of prominent Akron inventor Lewis Miller, became Edison's second wife in 1886.

Mina Miller was a brilliant and talented woman who could have achieved anything she envisioned. Her intent was to be a full partner and supporter of her husband's grand endeavors. Intrigued with Edison's work, Mina was frequently welcomed in Edison's West Orange laboratory, where she kept notebook records of several lighting experiments. She also used her abilities to create a beautiful home, care for the family, and to support Edison in the social aspects of his professional life. One of her most formidable tasks was integrating the three children from Edison's marriage to Mary Stillwell with her own growing family. This challenge was made more acute by the fact that Mina was, at age twenty, less than seven years older than Thomas's first child, Marion.

Mina and Thomas had three children together. Edison revered his wife. He bought her a mansion in West Orange, New Jersey, and his comment about the purchase encapsulates his feelings. "It's a great deal too nice for me, but it isn't half nice enough for my little wife." And he once scrawled a note in her gardening book, "Mina Miller Edison is the sweetest little woman who ever bestowed love on a miserable homely good for nothing male." Mina supported Thomas in all his endeavors, and prided herself on her skill as Edison's "domestic engineer." She said to reporters in 1925 that she had always "tried to organize our home and our home life to give results just as much as the laboratory." Mina nursed him as his health declined and toward the end, her voice was the only one to which he would respond.

EDISON'S RELIGION AND ETHICAL PHILOSOPHY

Edison's mother, his wife Mina, and his father-in-law Lewis were all deeply religious, and attended church regularly. But Edison did not have much interest in formal religion. He joked that that he was "oblivious of Sunday." Edison did, however, believe in a divine Creator. His notion of the Creator was one who had set forth precise laws of the universe that were to be known through science and mathematics. As he phrased it, "I know this world is ruled by infinite intelligence. Everything that surrounds us—everything that exists—proves that there are infinite laws behind it. There can be no denying this fact. It is mathematical in its precision." But Edison's God wasn't a personal one. As he wrote in his diary, "What a wonderfully small idea mankind has of the almighty! My impression is that he has made unchangeable laws to govern this and billions of other worlds, and that he has forgotten even the existence of this little mote of ours ages ago. Why can't man follow up and practice the teachings of his own conscience, mind his own business, and not obtrude his purposely created finite mind in affairs that will be attended to without any volunteer advice."

Edison's idea of aligning with those unchangeable "infinite laws" and following "the teachings of his own conscience" meant living by a moral code grounded in honesty, respect, fairness, and integrity. He felt that the highest standards of personal and business ethics were congruent with the precise design of the infinite intelligence. Moreover, Edison hoped that his innovations would help humanity evolve to a higher moral plane. He proclaimed, "The machine has been human being's most effective escape from bondage." Like Gandhi, he believed that "Nonviolence leads to the highest ethics, which is the goal of all evolution. Until we stop harming all other living beings, we are still savages." When he was asked to serve on the Naval Consulting Board during World War I, he made it clear that he would only work on defensive weaponry. As he noted, "I am proud of the fact that I never invented weapons to kill."

Edison's religious and ethical philosophy is probably best summarized by his observation that, "If we all try to carry out the Golden Rule in this life we have little to fear from the hereafter no matter what our belief may be."

Edison at age sixty-three.

EDISON'S FOLLIES AND FOIBLES

For Edison, there was no such thing as failure. He viewed all outcomes as fascinating opportunities for learning. Yet, having discussed some of his greatest achievements, it's useful to consider some of his weaknesses and to understand some of his endeavors that didn't turn out so well.

His iron ore–milling venture is one of his most notorious misjudgments, sometimes referred to as "Edison's Folly." It was a folly not just because he lost money but because of the enormous amount of time and energy he invested that might have been spent more productively elsewhere.

Edison's judgment was also askew when it came to choosing the music that was to be recorded in the early days of the phonograph. He insisted on being the ultimate arbiter of what was worthy of reproduction on the machine he referred to as "my baby." As one musician whom he employed commented about Edison's musicality, "He didn't have any." And one of his business associates warned, "A one man opinion on tunes is all wrong. Last year when you were the only picker of tunes, you refused to let us record the four biggest successes of the year."

Edison also lost his usual sense of objectivity and fairness in the debate over the efficacy and safety of alterating current (AC) vs. direct current (DC) electric power. Edison's rival George Westinghouse championed AC over Edison's preference for DC.

In an effort to discredit Westinghouse and highlight what he believed to be the dangers of AC, Edison's representatives used it to electrocute animals in public demonstrations. And, despite his opposition to capital punishment, Edison encouraged the State of New York to utilize AC power in its electric chair. Edison's attorney went so far as to propose that the chair be christened "The Westinghouse." The first use of the chair was meant to be a grand demonstration of the lethal nature of AC power, but the condemned man absorbed seventeen seconds' worth of full charge before one of the electrodes began to give off smoke. The current was stopped and the executioners believed their job to have been completed until they noticed that the victim was still breathing. So they fired up the current again, this time for seventy-two seconds, until the unfortunate prisoner's body

EDISON AND TESLA: POWER STRUGGLE

Nikola Tesla (1856–1943) was an extraordinary genius who did seminal work in establishing the effective functioning of electrical power systems that could transmit power over long distances. Born in Croatia of Serbian parents, Tesla learned six languages and received a superb education in mathematics, physics, and engineering before making his way to Paris in 1882, where he met Charles Batchelor. Tesla's understanding of and interest in electricity led Batchelor to recommend him to work in one of Edison's companies.

In 1884, Tesla traveled to the United States to begin working for the Edison Machine Works in New York City. Unfortunately, Tesla and Edison soon had a falling out as Tesla desired to pursue improvements to arc lighting technology, which Edison believed was passé due to his own incandescent lighting innovations. Tesla left the Edison Machine Works in 1885 to start his own company, and began producing motors and generators using alternating current (AC). Tesla applied for the first of his more than seven hundred patents—many of them related to AC—on May 6, 1885. On May 16, 1888, he presented a lecture to the American Institute of Electrical Engineers (AIEE) on "a new system of motors and transformers of alternate currents." Immediately thereafter, George Westinghouse—one of Edison's chief rivals—bought the rights to forty of Tesla's AC patents and the stage was set for a tremendous power struggle between the Edison and Westinghouse companies.

Although Tesla designed the world's first successful hydroelectric power plant and helped Westinghouse's AC standard prevail in his battle against Edison's DC, Tesla lost the longer term public relations battle with Edison. Edison and Tesla were phenomenal geniuses who slept little and were each driven by a compelling desire to contribute to a better world. But their styles were diametrically opposed. Edison reveled in literally getting his hands dirty while Tesla was an obsessive-compulsive germphobe. Edison was largely self-taught and rather contemptuous of academia; Tesla was the recipient of a first-class European education and felt that Edison relied too heavily on empiricism. Edison was a family man; Tesla remained a bachelor throughout his life. Edison was the embodiment of American culture and ideals; Tesla felt that European culture was one hundred years ahead of what he found in the United States. Edison was a consummate innovator; Tesla's creativity was not as focused on the practical business applications of his inventions.

Ultimately, although Tesla was correct about the advantages of AC, Edison gained the edge in the marketplace and public mind due to his ability to assemble a team of collaborators and develop a systematic approach to innovation. Tesla functioned primarily as a solo genius. The six-foot-four-inch Serbian had an uncanny ability to work out all the details of highly complex problems in his mind. He was truly a visionary, yet often lacked the ability to practically apply his inventions to the marketplace. For example, Tesla envisioned linking the world's telephone and telegraph services through wireless technology; then, he designed a tower in Long Island for this purpose. Tesla's tower would also be able to transmit news, stock, and weather reports; pictures; personal messages; and secure government and military communications. He spoke prophetically to financier J. P. Morgan, "When wireless is fully applied the earth will be converted into a huge brain, capable of response in every one of its parts." Morgan ultimately withdrew his sponsorship of Tesla's tower concept, however, because he didn't support Tesla's altruistic notion that power and communication be freely available to all. Morgan's classic line when he pulled the financial plug was, "If anyone can draw on the power, where do we put the meter?"

Despite rumors that flooded the press over many years, neither man was ever awarded a Nobel Prize for science; but, in another touch of irony, Tesla did receive the coveted "Edison Medal" from the American Institute of Electrical Engineers in 1917—an honor he did not refuse.

started burning from the inside. Newspapers around the world decried the whole gory phenomenon and Edison's role in it.

After several years, AC was proven to be superior to DC because it allowed the efficient use of transformers to step voltage up and down, which was the only practical way to transmit power over long distances.

Edison has also been criticized for engaging in ethnic and racial stereotypes and for his close friendship with Henry Ford, who was an anti-Semite and racist. But although Edison did engage in stereotyping, he ultimately viewed people by what they could do rather than by race or background. As is largely true in Silicon Valley today, his interest in achievement and results was stronger than his attachment to any stereotype.

Of course, Edison could be stubborn, overbearing, and at times he was ruthless in his business dealings. In the course of his career he made a number of significant financial and strategic blunders. Although Edison had a long fuse, when angered he was capable of verbally eviscerating anyone unlucky enough to be the object of his wrath. As one employee observed, "[Edison] could wither one with his biting sarcasm or ridicule one into extinction." He wasn't always the avuncular teddy bear genius that the media loved, and his occasional tongue lashings would not be acceptable in today's workplace. But, learning more about his flaws, misjudgments, and weaknesses adds depth and dimension to our appreciation of his gifts and achievements. And, as he always emphasized, he wasn't a "wizard" or a "magician," and he didn't even like to be called a "genius." He was human, and he believed that all people are capable of great achievements if they work hard and use their full creative potential.

EDISON'S COMMON TOUCH

> I have friends in overalls whose friendship I would not swap for the favor of the kings of the world.
> —*Thomas Edison*

Thomas Edison was the supreme embodiment of Yankee ingenuity. As one newspaper commented, "Mr. Edison is in many respects an odd man and in every way peculiarly American." What made him peculiarly American? His charismatic optimism, orientation to self-improvement, and passion for continuous innovation were perfect expressions of the strengths of the American psyche, and followed in the tradition of Franklin, Jefferson, and Lincoln. These qualities were complemented by an unpretentious demeanor, a willingness to work hard, and a great sense of play and humor.

Although he was able to move easily among different levels of society, Edison always managed to project an image as a man of the people. As one chronicler commented, "A man of common sense would feel at home with him in a minute; but a nob or prig would be sadly out of place." Reporters reveled in describing his unkempt appearance, portraying him in his chemical-splattered work clothes, hands dirty, hair uncombed, and eyes ablaze with the passion for invention.

Edison exuded a natural charm that one reporter described as an "affability and playful boyishness." Sigmund Freud wrote of Leonardo da Vinci that "he continued to play as a child throughout his adult life thus baffling his contemporaries." Einstein, Edison's contemporary, was also well known for his playful, childlike persona. Like these other great geniuses, Edison maintained an unpretentious, childlike openness and passion for learning throughout his life.

This quality endeared him to the public and made him seem accessible, human, and real. This playful attitude also manifested itself in his sense of humor. As Edison's contemporaneous biographers Frank Dyer and Thomas Martin describe it, "His sense of humor is intense, but not of the hothouse, overdeveloped variety." Edison's common touch and sense of humor led him to play all sorts of pranks. One of his favorites was to present himself at his company's legal department and apply for a job as an inventor. The embarrassment junior clerks felt

EDISON'S GREEN VISION

Thomas Edison's love of nature and his passion for efficiency translated into a practical concern for energy conservation and environmental protection. By 1910, Edison had developed a storage battery that could power automobiles, trucks, and machines. He hoped this development would lead to the use of batteries as a self-sufficient source of energy in homes and buildings. In 1912, he constructed and helped to create a model home in West Orange, New Jersey, that was "off the grid," and powered solely by his storage batteries. He also began thinking about ways to harness the power of the wind and sun. Shortly before his death in 1931, Edison told his friends Henry Ford and Harvey Firestone, "I'd put money on the sun and solar energy. What a source of power! I hope we don't have to wait until oil and coal run out before we tackle that."

Edison in 1915, walking on the deck of the U.S.S. E-2 submarine to review an installation of his storage battery in the engine room.

having failed to recognize their boss immediately turned to an acknowledgement of the great man as easygoing and approachable. On another occasion, when the West Orange porter's gate guard failed to recognize Edison as he returned from a meeting, Edison did not protest. Instead, he waited patiently until one of the laboratory staff arrived at the gate to identify him. Edison's unpretentious manner endeared him to his employees, adding to the aura and mystique of his extraordinary genius.

CULTURAL ICON, NATIONAL TREASURE

A few years before the West Orange laboratory reached its zenith during the First World War, Edison consolidated all his businesses into one. Thomas A. Edison Incorporated was established in 1911, with Edison as president and chairman. Now, at age sixty-four, he began to delegate more of the day-to-day running of the business to his trusted associates. Although he continued to develop and apply for patents until the last year of his life—he filed for his 1,093rd at age eighty-four—Edison's primary focus gradually shifted from driving new innovations that would radically change the world to the further consolidation and refinement of the world-changing inventions he had already developed. He also began to invest more time and energy in the consolidation of his legacy.

As World War I began, Edison was asked to become the director of the Naval Consulting Board, a think tank focused on reviewing and improving military technology. He was keen to contribute to defensive technologies and worked on methods of gun location and submarine detection. In the late 1920s, at the request of his friends Henry Ford and Harvey Firestone, Edison began work on a project to find alternatives to rubber for automobile tires. He experimented with thousands of substances and made significant advances. In 1928, Edison was awarded a special Congressional Medal of Honor, and 1929 marked the fifty-

LINCOLN AND EDISON:
GREATNESS, HUMILITY, PERSPECTIVE

Representative stanzas of the favorite poems of two great men express remarkably similar perspectives:

> Oh, why should the spirit of mortal be proud?
> Like a swift-fleeting meteor, a fast-flying cloud,
> A flash of the lightning, a break of the wave,
> He passes from life to his rest in the grave.
> —*From "Mortality" by William Knox,*
> *Lincoln's favorite poem*

> The boast of heraldry, the pomp of power,
> And all that beauty, all that wealth e'er gave,
> Awaits alike th' inevitable hour:
> The paths of glory lead but to the grave.
> —*From "Elegy in a Country Church Yard"*
> *by Thomas Gray, Edison's favorite poem*

year anniversary of the invention of the incandescent light bulb and the whole nation joined in honoring Edison's Golden Jubilee. Although his health was declining, his status as a living legend continued to grow. Luminaries from all walks of life, including President Herbert Hoover, Marie Curie, Charles Lindbergh, and Henry Ford came to pay tribute to the cultural icon who embodied American leadership in innovation.

As the summer of 1931 came to a close, Edison's incredible and seemingly inexhaustible life force began to seep away. He fell in and out of a coma, breathing his last breath in the early morning of October 18, 1931. After his memorial service and burial on October 22, President Hoover led the nation in commemorating the life of the man who had illuminated the world by asking citizens to turn off the lights in their homes for one minute. Radio networks throughout the land followed suit by going silent for the same minute in tribute to the man who did more than any other to create the technology for "such stuff as dreams are made on."

PART TWO

EDISON'S FIVE COMPETENCIES OF INNOVATION

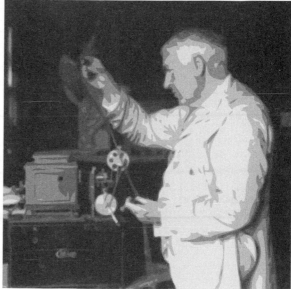

Chapter Three

COMPETENCY #1— SOLUTION-CENTERED MINDSET

The word "solution" comes from the root *solvere*, which means "to loosen or set free." A solution-centered mindset is the first step in setting free your Edisonian powers. Edison took on big problems throughout his career, but he always approached them with the attitude that his success was inevitable. And this became a self-fulfilling prophecy that brought him fulfillment and wealth as it changed the world. Edison knew that the word "problem" comes from the roots *pro*, meaning "forward," and *ballein*, meaning to "throw or drive." The bigger the problem, the greater the challenge, the more determined Edison became to "drive forward." Once Edison identified an objective he wished to achieve, he pursued it passionately and with optimistic resolve. Undeterred by formidable obstacles and seeming failures, he was not discouraged when results proved elusive.

Edison's astounding ability to innovate began with his process of mentally preparing for the "hunt," a term he and his laboratory staff used to describe their intensive search for solutions. What are the critical elements of Edison's winning attitude—his *solution-centered mindset*—and how can you cultivate them for yourself and your organization?

We will provide the answer as we guide you through the first competency for *Innovating Like Edison*. It comprises the following five elements:

1. Align Your Goals with Your Passions
2. Cultivate Charismatic Optimism
3. Seek Knowledge Relentlessly

4. Experiment Persistently

5. Pursue Rigorous Objectivity

ELEMENT 1: ALIGN YOUR GOALS WITH YOUR PASSIONS

Cherish your visions and your dreams, as they are the children of your soul, the blueprints of your ultimate achievements.
—*Napoleon Hill*

A recent Google search on the topic of goal setting yielded more than twenty million links. There are a myriad of ways to talk about setting goals, but everything you really need to know is present in Edison's approach. As we've mentioned, Edison was a prime inspiration for Napoleon Hill's *Think and Grow Rich*. Hill states that harmony and success require that "one must marry one's feelings to one's beliefs and ideas." Hill adds, "Desire is the starting point of all achievement, not a hope, not a wish, but a keen pulsating desire which transcends everything."

From his earliest days as an amateur chemist to the last years of his life when he was revered internationally as an icon of genius, Edison's daily activities reflected the alignment of his goals with his passions. Edison's feelings were wed to his ideas and his remarkable achievements were driven by that "keen pulsating desire which transcends everything." The marriage of his goals and passions allowed Edison to savor the process of achievement as much the result. As he commented, "I never did a day's work in my life, it was all fun."

Although he realized early in his career that commercial success was essential to maintaining the opportunity to do what he loved, Edison was never motivated by specific financial goals. He attained great wealth, but the riches he accrued were merely an epiphenomenon. As he expressed it, "One might think that the money value of an invention constitutes its reward to the man who loves his work. But . . . I continue to find my greatest pleasure, and so my reward, in the work that precedes what the world calls success."

Edison propelled himself from within; first, from his deeply rooted desire to pursue knowledge, and then, to provide products and services that would improve the quality of people's lives.

Edison's passionate desire to "surprise Nature into revealing her secrets" was evident in his childhood and continued unabated until his last breath. From his quest to own his first chemistry set, then build a working telegraph line while formulating plans to become a master telegrapher, to his dream of making the phonograph available to every household and lighting the world, Edison was always able to frame his goals clearly. He knew exactly what he wanted to achieve and he was able to vividly envision the reality of the solutions he sought. Because his goals were always expressions of his passion, he savored every step in the process of their realization.

Contemporary psychological research validates Edison's approach, and supports the notion that we can all learn how to develop this essential element of success. Dr. John Dacey, professor emeritus of developmental psychology at Boston College, and Dr. Kathleen Lennon of Framingham State College studied and then condensed decades of research into what makes scientists, writers, business leaders, musicians, and other powerfully creative individuals successful. Among the most important traits they uncovered are: 1) passionate goal directedness, and 2) perseverance through self-control.

Dacey and Lennon emphasize that both of those qualities can be developed by adults even if they do not possess those traits in younger years, as Edison clearly did. Citing numerous studies by psychologists including Mihaly Csikszentmihalyi, Paul Torrance, David Perkins, Robert Weber, and others, Dacey and Lennon conclude that passionate goal directedness helps successful individuals generate "great amounts of energy to invest intensely in their work." These goals are typically long term in nature and associated with a big vision; so the second skill, perseverance through self-control, is critical in allowing fulfillment of the first. Dacey and Lennon define self-control as an individual's willingness to "persevere in the face of frustration."

In other words, success is a function of perseverance, and perseverance is driven by aligning passions with big, long-term goals. Edison's success was the result of his "passionate goal directedness." His "pulsating desire" allowed him to "transcend everything" so that frustrations, obstacles, and difficulties seemed to provide him with even more energy. As a colleague remarked, "Edison seemed pleased when he used to run up against a serious difficulty. It would seem to stiffen his backbone and make him more prolific of new ideas. For a time, I thought it was foolish to imagine such a thing, but I could never get away from the impression that he really appeared happy when he ran up against a serious snag."

Dr. Richard Restak, a clinical professor of neuroscience at the George Washington University Hospital School of Medical and Health Sciences, offers further validation of Edison's approach. Restak argues that many goals go unfulfilled or are prematurely abandoned because they are not designed robustly enough to mobilize the brain. He points out that for the brain to remember to organize behavior in alignment with a goal it **must connect the emotional component with its rational component**. This alignment links the prefrontal cortex with the limbic system, thereby dramatically enhancing the likelihood that the goal will be remembered and translated into behavior.

Understanding how to set goals so that they will be remembered and translated into behavior is a simple, critical key to successful innovation and, of course, to personal happiness and fulfillment. Yet despite the wealth of information available on the topic, most organizational innovation efforts fail because they don't define their goals clearly, and they neglect to align goals with emotions. Innovation literacy begins with a practical understanding of how to define and align your personal goals.

Creating Innovation Literacy: Align Your Goals with Your Passions

Edison's ability to manifest his goals was predicated on his intuitive understanding of how the creative mind functions. You can access and apply the same principles that he did; and, to make it easy, we've consolidated everything you need to know about goal manifestation in an acronym: SMART EDISON.

Many people are familiar with the SMART acronym but few apply it. The EDISON makes the SMART part come to life. Start with SMART:

S—*Specific*: Define exactly what you want to accomplish in detail.

M—*Measurable*: Decide how you will measure your progress. How will you know that you've achieved your goal?

A—*Accountable*: Make a full commitment to be personally responsible for achieving your goal. When setting goals in a team, be certain that accountability is clear.

R—*Relevant*: Ensure that your goals are relevant to your overall purpose and values.

T—*Timeline*: Create a clear timeline for the achievement of your goals. As Napoleon Hill puts it, "A goal is a dream with a deadline."

Then, invoke EDISON:

E—*Emotional:* As the research of Restak and others demonstrates and the story of Edison confirms, the emotional element of goal manifestation is crucial. Express your goal in terms that energize and excite you. Feel the passion associated with the fulfillment of your goal.

D—*Decisive:* Avoid wishing or hoping for the realization of your goal. Instead, just decide with the full force of your being to make it so. It's essential to make a committed decision even if you can't yet see the path forward to the realization of your goal.

I—*Integrated:* Please be careful what you decide to manifest. It's important to consider how each goal you call into being will be integrated with your other goals and your overall purpose. (Hint: Always link your goals to a higher purpose beyond your own personal benefit.)

S—*Sensory:* Use all your senses to vividly imagine the manifestation of your goal. Speak your goal. Draw it. Dance it. Taste it.

O—*Optimistic:* Approach the process of goal manifestation as if the process works like gravity. Act as if it is simply a universal law that will work every time.

N—*Now:* Envision and express your goal in present-centered terms. And begin it now!

Steve Odland, chairman and CEO of Office Depot, enthuses about the alignment of goals and passions in his organization:

I think that goal-setting is an important point, not so much from setting a numerical goal as creating possibilities for people. I think people are often limited by their own interpretation of what they think is—or isn't—possible. In the case of company performance from a financial standpoint, people many times are corralled by expectations set externally. So the whole notion of external targets set by public companies is one that sometimes works backwards on employees. Employees will say, "Well, our target is to deliver 12 percent earnings per share," and then everybody works toward 12 percent earnings. Instead, I want my people to be reaching beyond that, and thinking about what is possible rather than solving to a 12 percent goal.

We've created a vision statement that helps us do this. It gives people a destination and a compass. We get them thinking about solutions rather than specific products or services. You can't limit your goal statements to rational descriptions . . . You have to be irrational about it because all invention requires irrationality. You have to capture people's imagination through inspiration, and I know Edison understood that in his work. In our case at Office Depot we have fifty thousand people. Inspirational goals get everybody thinking about the possibilities of invention, the possibilities of something different, and coming together to create it.

Odland adds, "The only difference between 'possible' and 'impossible' is an idea, isn't it?"

> All the breaks you need in life wait within your imagination. Imagination is the workshop of your mind, capable of turning mind energy into accomplishment and wealth.
> —Napoleon Hill

ELEMENT 2: CULTIVATE CHARISMATIC OPTIMISM

Thalia, Aglaia, and Euphrosyne were the three graces of Greek mythology, also known as the three charities. The root *chari* in the words "charities" and "charismatic" refers to grace or divine favor. Edison was graced with a natural optimism. As Dyer and Martin express it, "Optimistic and hopeful to a high degree, Edison had the happy faculty of beginning the day as open-minded as a child—yesterday's disappointments and failures discarded and discounted by the alluring possibilities of tomorrow."

His outlook was disarmingly simple. He stated it this way: ". . . look on the bright side of everything." Edison found the silver lining in every cloud, a skill he developed while observing his father, Samuel, persist through many life challenges. As a child, Edison observed how his father—a political activist, innkeeper, farmer, and grain merchant—was forced to continually reinvent his career in the face of steep economic challenges, including those brought on by the Civil War. Samuel wore many hats over the course of Thomas's first fifteen years just to remain solvent. Despite his many difficulties, Edison's father maintained, "A lively disposition always looking on the bright side of things . . . full of most sanguine speculation as to any project he takes in his head."

In the laboratory, Edison's coworkers often remarked that he approached every project with a demeanor of "happiness," a sense of eternal freshness that allowed him to maintain an upbeat view even when difficulties arose. Dr. E. G. Acheson, an experimenter once employed on Edison's staff, summed it up this way:

In the summer of 1926, Edison delivers an uplifting radio address to the nation while sitting in the rear garden of the Miller family's vacation home in Chautauqua, New York.

> I once made an experiment in Edison's laboratory at Menlo Park during the latter part of 1880, and the results were not as looked for. I considered the experiment a perfect failure, and while bemoaning the results of this apparent failure Mr. Edison entered, and, after learning the facts of the case, cheerfully remarked that I should not look upon it as a failure, for he considered every experiment a success, as in all cases it cleared up the atmosphere, and even though it failed to accomplish the results sought for, it should prove a valuable lesson for guidance in future work. I believe that Mr. Edison's success as an experimenter was, to a large extent, due to this happy view of all experiments.

Edison's "happy view of all experiments" helped him persist in the face of adversity, resisting the temptation to call an "unexpected outcome" a failure. He believed that most people gave up too soon, walking away from success and accepting failure instead. He said, "Our greatest weakness lies in giving up. The most certain way to succeed is always to try just one more time."

Even when a situation appeared catastrophic, Edison remained optimistic. In 1914, when Edison was sixty-seven, he watched calmly while a raging fire burned six new phonograph factory buildings to the ground and gutted seven others at the West Orange complex. A small wooden shed holding inflammable motion picture film somehow had caught fire, casting flames onto the original wooden Phonograph Works building, and then spreading rapidly to several surrounding structures made of cement—a material believed to be fireproof in the early years of the twentieth century. How did Edison respond? He proclaimed that he would resume manufacturing phonograph records within ten days and began mapping out his rebuilding campaign immediately.

Estimates placed Edison's loss in physical structures, master phonograph recordings, manufacturing equipment, and other valuables as high as $7 million—the equivalent of $106 million today. The buildings were covered by $2 million in regular insurance, with the balance covered by self-insurance under the Edison Company umbrella, requiring Edison to make up the difference from his own pocket. His son Charles, age twenty-two, ran to his father's side, thinking Edison would be devastated. Instead, Charles was amazed to see his father smiling and bidding him to run and get Mina, saying that she'd never have a chance to see anything like this again in her entire life.

Edison's unflinching response to the fire highlights the deeply embedded nature of his optimism. As Paul Israel describes it, "Where others might see disaster and failure he was always optimistically looking for opportunities and seeing the possibility of new directions for improvements." Edison spoke of rebuilding the phonograph factories in a manner that "took advantage of the latest improvements in factory design developed by Henry Ford," who was a leader in modern factory design at that time. Edison said, "I am repairing my concrete buildings and wish to arrange my machinery properly in order to take advantage of Mr. Ford's methods as far as possible."

Edison's optimism created an irresistible magnetism that drew others toward him. He inspired confidence in colleagues, customers, vendors, journalists, and investors. Even when circumstances appeared dire, Edison was able to get people to give him the benefit of the doubt. For example, after working for nearly a year on the storage battery, Edison's experiments had yielded not one single expected outcome. When asked by associate Walter S. Mallory why he persisted in working to invent a new storage battery even though he was getting no results, Edison cheerily replied, "Results! Why, man, I have gotten a lot of results. I know several thousand things that won't work!" Edison viewed negative events as temporary setbacks on his inevitable path to success.

The word "charismatic" means "possessing an extraordinary ability to attract." Edison's charismatic optimism gave him an extraordinary ability to attract support for his innovations. His positive attitude energized his financial supporters so he could continue to afford to inspire his staff and customers. As Daniel Craig, one of his investors, commented, "Your notes, like your confident face, always inspire us with new vim."

As Edison became a national figure, his positive outlook had an even broader effect. He helped encourage the nation through tough times. People from all walks of life were uplifted when he said, "Be courageous. I have seen many depressions in business. Always America has emerged from these stronger and more prosperous. Be brave as your fathers before you. Have faith! Go forward!"

Edison knew the truth expressed in the famous line from Shakespeare, his favorite playwright: "Our doubts are traitors and make us lose the good we oft might gain by fearing to attempt." Edison's optimism drove him to attempt and achieve his world-changing innovations. Psychol-

Edison stands next to one of the cars that were driven up steep hills to show the power and durability of the improved Edison Storage Battery.

ogist Karen Horney (1885–1952) undertook groundbreaking work on optimism early in the twentieth century. She discovered that most people actually succeed when they decide, with full commitment, to accomplish something. Most of what people label as "failure," according to her research, is a function of their doubts acting as traitors, causing them to withhold full commitment and give up too soon. As Edison phrased it, "Nearly every man who develops an idea works it up to the point where it looks impossible, and then he gets discouraged. That's not the place to become discouraged." He adds, "Many of life's failures are people who did not realize how close they were to success when they gave up."

Dr. Martin Seligman, director of the University of Pennsylvania's Positive Psychology Center, and author of *Learned Optimism*, has researched optimism for more than thirty years. His work offers further confirmation of the validity of Edison's approach. According to Seligman, optimists get better results than pessimists in most areas of life. His research shows that optimists perform better at school, in relationships, on the athletic field, and at work. Their resistance to colds and other illnesses is stronger, they have longer-lasting, happier relationships, and they recover faster from illness and injury. And optimists make significantly more money. All this is true despite the fact that pessimists are more skilled, according to Seligman, in their ability to analyze current problems accurately. For the pessimist, the optimist is someone who simply doesn't yet see the facts as they are. But despite the tendency to view the world through rose-colored glasses, numerous long-term studies confirm that better results emerge when we err on the side of optimism.

When pessimists read the paragraphs above, they tend to respond: "Oh, great. I'm destined to underperform in all areas of life. I knew my situation was hopeless." The good news is that the situation isn't hopeless because, as Seligman emphasizes, optimism is something you can learn. It's an essential element of your personal innovation literacy.

Creating Innovation Literacy: Cultivate Charismatic Optimism

> Every adversity, every failure, every heartache carries with it the seed of an equal or greater benefit.
> —Napoleon Hill

There is no innovation without optimism. As Daniel Goleman emphasizes in his excellent book *Working with Emotional Intelligence*, optimism is an important aspect of emotional intelligence and a critical characteristic for leadership. The good news is that the work of Goleman, Seligman, and others demonstrates that this critical element of Edison's approach is something that you can develop.

You can improve your chances for success in every area of life, especially innovation, by cultivating what Seligman calls an optimistic "explanatory style." Explanatory style refers to the way that you "coach" yourself, particularly in the face of obstacles and adversity. In the face of apparent misfortune, adversity, or bad news, pessimists focus on the negative, and then take it *personally* ("It's all my fault"), assume the difficulty is *permanent* ("It will never improve"), and consider its effect to be *pervasive* ("It will totally ruin my life"). This sort of interior monologue locks in pessimistic conclusions, so that they become self-fulfilling prophecies.

When optimists confront challenging circumstances, they respond differently. They can see the influence of external factors in their problems so they *don't take it personally*. Optimists expect success and consider happiness to be their normal state so they view negative events as *temporary glitches* on their path to inevitable progress. Moreover, when something goes wrong the optimist views it as an *isolated phenomenon*, insulated from other aspects of their lives.

Optimists and pessimists reverse this self-coaching pattern when good things come their way. While optimists believe that they "make their own luck," pessimists feel that they just "got lucky." Optimists feel that their good fortune is natural and destined to continue, pessimists are sure that it's only temporary. And optimists know that anything good that happens will improve their whole lives, while pessimists maintain their focus on all the aspects of their lives that are unsatisfactory.

The solution to pessimistic self-coaching is simple. You can liberate yourself from the self-defeating explanatory style of the pessimists by consciously choosing and practicing a more optimistic approach to self-coaching. If you imagine, for example, that you've spent considerable time and energy researching and developing a proposal for an innovative product or service and you've finally gotten the chance to present it to your company's senior team or board, and they respond with a flat "no."

The pessimists respond to this rejection by personalizing: "It's my fault. My proposal isn't good enough and neither am I"; assuming the failure is permanent: "I blew it. They'll never give me another opportunity"; and concluding that the damage is pervasive: "I'm a failure. My whole life is worthless."

The optimists respond with a much different and more adaptive inner dialogue. Instead of blaming themselves for the rejection, the optimists think, "The members of this board aren't receptive to innovation; I will have to think of a stronger way to get through to them." Instead of assuming that rejection is permanent, optimists focus on a better future: "New board members are elected next quarter. I will try again then. Meantime, maybe I can find a venture capital firm to invest in my idea. In any case, I will work out the glitches in my presentation so that it will be impossible to resist." And, rather than concluding that the damage is pervasive, the optimists decide, "This is a great learning experience; it will help me improve everything I do."

The important thing to understand about explanatory style is that it is something **you can choose to change**, even if your habitual response to negative events is to interpret them pessimistically. When faced with challenging circumstances, you can internalize the power of charismatic optimism by asking yourself: How would Edison respond? How would he interpret these events?

Edison's example, in both action and spirit, can serve as a supercharging dynamo, energizing your mind's success circuitry. Edison understood intuitively that the human brain is the most profoundly powerful learning and creative problem-solving mechanism in the known universe. He knew that we are hardwired for success; and as Dr. Seligman along with other researchers would later prove, that pessimism "short-circuits" the success mechanism. Optimists get better results, not because positive thinking summons New Age fairies to sprinkle happy dust on their efforts, but rather because they remain engaged and **passionately focused on finding solutions**.

Three of Edison's fellow inductees of the National Inventors Hall of Fame express this element of their *solution-centered mindset:*

It just never occurred to us that this might be an unsolvable problem.
    —*Dr. Donald Keck, co-inventor of optical fiber*

I'm always enthusiastic. I don't believe there's anything that's impossible.
    —*Dr. Jim West, co-inventor of the electret microphone*

So many times when you try to do something in science, when you try to invent something, people tell you that it's impossible, that it will never work. But I think that is very rarely true. I think that if you really believe in yourself, that if you really stick to things, there is very little that is really impossible.
    —*Dr. Robert Langer, inventor of systems for*
    controlled release of macromolecules

ELEMENT 3: SEEK KNOWLEDGE RELENTLESSLY

Every child is curious, but the young Edison was supercurious. His insatiable thirst to understand the world manifested in an endless stream of questions. As soon as he could speak he began asking about *everything* in an intensely penetrating, focused, and comprehensive manner. His relentlessness, combined with his tendency to ask about things that seemed obvious to others, made his family wonder if there was something awry. As Dyer and Martin note:

His questions were so ceaseless and innumerable that the penetrating curiosity of an unusually strong mind was regarded as deficiency in powers of comprehension, and the father himself, a man of no mean ingenuity and ability, reports that the child, although capable of reducing him to exhaustion by endless inquiries, was often spoken of as rather wanting in ordinary acumen.

Edison spent less than three months in the local Port Huron school system. As he recalled, "I used never to be able to get along at school. I don't know what it was, but I was always at the foot of the class. I used to feel that the teachers never sympathized with me and that my father thought I was stupid, and at last I almost decided that I really must be a dunce."

Fortunately, Nancy Edison took over and channeled young Thomas's precocious mind toward the world of books. Edison fondly recalled her influence:

> I was a careless boy, and with a mother of different caliber I should have probably turned out badly. But her firmness, her sweetness, her goodness, were potent powers to keep me in the right path . . . My mother taught me how to read good books quickly and correctly, and as this opened up a great world in literature, I have always been very thankful for this early training.

It was through reading that Edison discovered his passion for science, especially chemistry. Dyer and Martin write of Edison's love of reading, cultivated during his homeschooling days: "Certain it is that under this simple regime studious habits were formed and a taste for literature developed that have lasted to this day. If ever there was a man who tore the heart out of books it is Edison, and what has once been read by him is never forgotten if useful or worthy of submission to the test of experiment."

With his mother's guidance, Edison discovered that he learned best by complementing his reading with hands-on experimentation. He later noted, "I was never able to make a fact my own without seeing it, and the description of the best works altogether failed to convey to my mind such knowledge of things as to allow myself to form a judgment upon them."

Edison "made facts his own" by touching and examining objects associated with his readings—such as ores, rocks, powders, leaves, wires, wood, bark, and other elements of nature. He intuitively understood the importance of experiential learning, combining multisensory information with words.

> To invent, you need a good imagination and a pile of junk.
> —*Thomas Edison*

When Edison desired to learn something, he wanted to do so as rapidly as possible. He taught himself how to speed-read, noting:

Still sharp at age eighty-three, Edison conducted a wide-ranging series of experiments with goldenrod. Although Edison proved that goldenrod yielded a natural source of highly elastic, rot-resistant rubber, he never launched it commercially.

After I became a telegraph operator, I practiced for a long time to become a rapid reader of print, and got so expert I could sense the meaning of a whole line at once. This faculty, I believe, should be taught in schools, as it appears to be easily acquired. Then one can read two or three books in a day, whereas if each word at a time only is sensed, reading is laborious.

His speed-reading skills allowed him to keep pace with the torrent of questions that poured through his mind on a daily basis. He read books on a remarkable range of subjects to address his endless queries. As Edison noted, "I didn't read a few books, I read the library."

Edison believed that reading was a key to self-improvement. He used reading as a means to bootstrap his way to new knowledge in the areas that supported his goals. His voracious reading, for example, gave him an advantage in his quest to become a master telegrapher. Edison devoured articles in the influential industry journal *The Telegrapher*, as well as other trade publications. The knowledge he gained through this independent study allowed Edison to accelerate his progress and achieve his goal quickly. He believed that reading would accelerate his development of new ideas, breakthrough solutions, original inventions, and that, ultimately, it would lead him to world-changing innovations.

Edison's signature style as an experimenter and inventor was deeply informed by the depth and breadth of his reading. He never began a round of experiments without first reading everything available on the subjects of his studies. Information gleaned from reading allowed Edison to establish diverse contexts for his hypotheses. He once noted this to a reporter during an interview in his cavernous office in the West Orange laboratory, in which he had a library stocked with more than ten thousand volumes: "When I want to discover something, I begin by reading up everything that has been done along that line in the past—that's what all these books in the library are for."

Through reading, Edison "cross-trained" himself in multiple disciplines, using books as a pathway into new fields of endeavor. Dyer and Martin describe the content of one major area of Edison's library:

Here may be found the popular magazines, together with those of a technical nature relating to electricity, chemistry, engineering, mechanics, building, cement, building materials, drugs, water and gas, power, automobiles, railroads, aeronautics, philosophy, hygiene, physics, telegraphy, mining, metallurgy, metals, music, and others; also theatrical weeklies as well as the proceedings and transactions of various learned and technical societies.

Edison intentionally designed his library to complement his learning process: reading first, then tangibly "experiencing" the compounds and substances relevant to his inquiry: "The [library] shelves are . . . filled with countless thousands of specimens of ores and minerals of every conceivable kind gathered from all parts of the world, and all tagged and numbered." Through physically touching a diverse assortment of these compounds, Edison engaged all his senses, examining structure and color, fragrance or odor, and even lightly tasting ores and rocks. Complementing his intensive reading with this kind of multisensory exploration, Edison cultivated an ability to understand the essential properties of the substances he studied. By structuring his library in this unique way, Edison continually reinforced the *solution-centered mindset* he needed to solve complex problems. As one of his long time associates remarked:

> . . . in addition to the knowledge he has acquired from books and observation, he appears to have an intuitive apprehension of the general order of things, as they might be supposed to exist in natural relation to each other. It has always seemed to me that he goes to the core of things at once.

Edison's intense knowledge-gathering abilities and his powerful approach to multisensory learning fostered an uncanny ability to accurately predict the outcomes of his experiments. Often, Edison's first guess at a result represented the ultimate answer. As Francis Upton, one of his close associates, observed: "One of the main impressions left upon me, after knowing Mr. Edison for many years, is the marvelous accuracy of his guesses. He will see the general nature of a result long before it can be reached by mathematical calculation."

Even at his in Fort Myers, Florida, retreat—which he purchased in the late 1890s as an escape from the intensities of West Orange—Edison maintained a vast library and planted world-class research gardens with nearly nine hundred exotic specimens of trees, flowers, and plants from all over the world. Thus, even while "loafing"—as he called it—on vacation, Edison continued his relentless pursuit of knowledge.

Creating Innovation Literacy: Seek Knowledge Relentlessly

You can take an important step in creating innovation literacy by understanding how to do what Edison did. Let's begin by considering the issue of learning style. Individuals tend to have a preference for how they process information. Some people learn best by seeing, others by listening, and some need to engage physically. Visual learners like to see things in writing. They enjoy daydreaming, doodling, and visualization. Auditory learners prefer to receive information through the medium of sound. They tend be good listeners. Kinesthetic learners like to learn by doing. They enjoy a hands-on approach.

Many of us were raised as if there were only two styles of learning: smart, and dumb. But it turns out that much of what was considered dumb is a different kind of smart. Edison's teacher at the Port Huron school probably concluded that his brain was "addled" because he wasn't receptive to auditory information. In other words, he didn't like to listen. Edison was a strong visual/kinesthetic learner. He loved to picture things in his mind's eye, which probably led his teacher to scold him for daydreaming, and he had a natural passion to physically explore the world around him, which meant he had no interest in sitting still.

Edison's strong visual modality led him to read voraciously; and his vivid kinesthetic preference inspired him to, whenever possible, physically explore his subjects. His knack for using his kinesthetic intelligence to aid in problem solving is manifest in his comment that, "Great ideas originate in the muscles."

It's a delightful irony that Edison used his strong visual/kinesthetic style to create two of the greatest innovations in audio: the carbon-button transmitter that dramatically improved the sound quality of Bell's telephone and, of course, the phonograph. To fully explore their properties, Edison often held telegraph and telephone equipment in his mouth and placed his face up against his prototype phonographs so he could feel the sound vibrations, taking advantage of his kinesthetic sense. He then employed his imagination to visualize and sketch additional ideas or improvements for these inventions.

If, unlike Edison, you are a strong auditory learner, then you will do best with books on tape. If you have a predominantly kinesthetic profile, then you will want to get involved in "action-learning" and hands-on experimentation. If you have a preference for visual learning—and we will assume that you have some degree of comfort with this modality since you are reading this book—then the best way to accelerate your learning process is to do what Edison did, and learn to speed-read.

Woody Allen described his experience of taking a speed-reading course by proclaiming, "We read *War and Peace* in an hour. It's about Russia." Despite the filmmaker's droll review of his experience, speed-reading—with improved comprehension—is something you can learn. Philosopher John Stuart Mill, presidents Franklin D. Roosevelt and John F. Kennedy and, of course, Thomas Edison all trained themselves to read more than one thousand words per minute.

Reading faster with greater comprehension can be approached in two complementary ways: 1) training your eyes and mind to take in larger groups of words in shorter periods of time, and 2) learning the most efficient and effective means of study.

You can train your eyes and mind to take in larger groups of words in shorter periods of time by practicing the exercises in Tony Buzan's classic *The Speed Reading Book*, or by attending one of his seminars. Buzan emphasizes that "the human visual system can photograph an entire page of print in one-twentieth of a second." Buzan offers a series of simple, progressive exercises for shortening the time needed to mentally photograph the words on a page. Paul Scheele offers another excellent approach in his PhotoReading programs, based on an integration of neurolinguistic programming and accelerated learning techniques. According to Scheele, proper training allows the learner to "bypass" the processing of the conscious mind, which is limited to seven bits of information at a time. Instead, PhotoReading emphasizes the "preconscious processor," which can absorb thousands of pieces of information simultaneously.

> One must be an inventor to read well. There is then creative reading as well as creative writing.
> —*Ralph Waldo Emerson*

Edison's mother guided him to apply the most effective and efficient means of study so that he could "tear the heart out of books." Here is how you can do what he did:

- *Set clear objectives*: By approaching all your reading with a clear idea of what you want to learn, you will learn faster and more effectively. In a classic study, two groups were asked to read the same book. The first group was told that they were "responsible for the whole book." The second group was given the objective of discerning only the book's three major themes. When tested, the "three themes" group did better on all aspects of the exam, including questions that were unrelated to the three themes. Formulate specific objectives for your reading and write them down.

- *Warm up your brain*: Tune in to your subject matter by expressing verbally—or on paper—what you already know about the subject. This will energize the associative network in your brain. In two or three minutes you can access your knowledge base, and by bringing it to the fore, you greatly improve the chances that you will actually remember what you read.

- *Overview the text*: Read the Contents page, introduction, exercises, chapter summaries, and review or conclusion sections first. Just like completing a jigsaw puzzle, you begin with the borders and work your way in. In many cases you will discover that your objectives have been achieved via the overview.

- *Record and share key points*: If you've set your objectives, warmed up your brain, and overviewed your book, you will have an excellent idea of where the remaining "gold" is hidden in the text. "Tearing the heart out of books" involves focusing in on the parts that are most relevant to your objectives and choosing to skip the rest. As Edison did, record the key points of your reading. After you've finished, teach someone what you learned. As you practice expressing what you've learned you'll consolidate your ability to recall the material and you'll gain a clearer idea what you still need to learn.

Of course, if Edison were alive today he would certainly be complementing his voracious reading with intensive research on the Internet. The Web offers an unprecedented cornucopia of knowledge, but it also overflows with drivel. The same basic principles—especially setting clear objectives—apply to getting the most from your explorations in cyberspace.

> I see what has been accomplished at great labor and expense in the past. I gather the data of many thousands of experts as a starting point and then I make thousands more.
> —*Thomas Edison*

TRIZ

Another contemporary research and problem-solving tool that Edison would have loved was developed by Genrich Altshuller (1926–1998), a former patent examiner for the Russian navy. Known as TRIZ (pronounced "treez," from the Cyrillic acronym, for Theory of Inventive Problem Solving), it is based on an intensive survey and analysis of the fundamental inventive principles underlying the creation of a wide diversity of patents. Altshuller, a gifted inventor in his own right, identified forty principles of invention and seventy-six standard solutions that can be applied through a disciplined analytical process. Just as modern chess computers bring together the combined wisdom of generations of grand masters, TRIZ brings together the most effective strategies of many of the world's great inventors.

TRIZ is based on the notion that someone, somewhere has probably already solved your problem or one like it. By cross-referencing solutions to many different types of problems, Altshuller and his associates have discovered patterns that are repeated across different disciplines, industries, and sciences. The 40 Inventive Principles provide a systematic, logical progression for the process of creative problem solving. For example, the first of the 40 Inventive Principles is Segmentation. Segmentation means dividing an object into independent parts, or making it modular/sectional. Examples of products that are expressions of this principle include attachable garden hoses, modular furniture, and flexible computer components. Altshuller's seventh principle is Nesting. Like sets of hollow wooden Russian dolls stacked inside each other, the nesting principle focuses on objects that contain other objects. Practical examples include stackable lawn chairs and telescoping antennas.

TRIZ has primarily been applied to making quality improvements for existing products in a wide range of industries. Many organizations use it in conjunction with their Six Sigma programs. (Introduced by Motorola in 1986, Six Sigma is a methodology for eliminating defects and improving quality.) TRIZ has been used to generate advances in technical fields like computer software development, chemical engineering, and architecture. Recently, there has been a growing exploration of TRIZ principles in "softer" disciplines such as education, marketing, and conflict management. Air Products, Boeing, Hasbro, Hewlett-Packard, Johnson & Johnson, Motorola, Samsung, Xerox, and many other innovative organizations have begun experimenting with this powerful tool.

> Dr. Jim West, a 1999 inductee into the National Inventors Hall of Fame, on seeking knowledge relentlessly:
>
> "I go through the literature and say, 'Boy, I hope nobody else has found this!' If that turns out to be true—if you can't find it in the literature—then I begin to feel good. And I try to push that envelope a little bit further."

Of course, as you learn more you realize that there is so much more to learn. As Edison phrased it, "We don't know a millionth of one percent about anything." The vastness of the unknown only served to inspire Edison's relentless quest for knowledge. This quest is an essential element of the *solution-centered mindset* that makes innovation possible.

ELEMENT 4: EXPERIMENT PERSISTENTLY

> Patience, persistence and perspiration make an unbeatable combination for success.
> —*Napoleon Hill*

The word "experiment" comes from the same root as the word "experience." An experiment is a procedure for testing the validity of a principle or hypothesis. It is a more formal, disciplined way to learn from our experience. A scientific experiment is the most formal and rigorous way of learning from experience.

For Edison, experimentation was the practical driving force of the innovation process. Grounded in the highest scientific standards of his day, Edison's trademark approach to experimentation began with the generation of a diverse array of hypotheses. He cast a wide net based on his reading and kinesthetic explorations, complemented by insights recorded in his notebooks.

Edison's prowess as an experimenter fed his confidence that he could "outinvent" any of his competitors. He believed that his dedication to preparing creatively for every experiment gave him a profound competitive advantage. Edison expressed his zeal in these words: "The only way to keep ahead of the procession is to experiment. If you don't, the other fellow will. When there's no experimenting, there's no progress. Stop experimenting and you go backward. If anything goes wrong, experiment until you get to the very bottom of the trouble."

Edison viewed every experiment he conducted as an important and meaningful step toward expanding the boundaries of his knowledge. In Edison's approach, failure didn't exist, because all outcomes yielded further data that he believed would ultimately provide solutions. His unshakeable belief that a solution to his questions existed drove his uncanny persistence in experimentation—and his persistence made his belief come true.

In 1915, Richard McLaurin, president of the Massachusetts Institute of Technology, delivered a speech on "Mr. Edison's Service for Science." He stated: "Edison has proved himself a great force . . . by giving so brilliant an exhibition of the method of science, the method of experimentation. When we get to the root of the matter we see that nearly all great advances are made by improvement in method."

Edison's mastery of experimental methods was manifest in his establishment of the world's first Research and Development laboratory at Menlo Park in 1876, and just over ten years later, the birthplace of Industrial Research and Development at his laboratory in West Orange, New Jersey. By creating these state-of-the-art facilities, Edison's laid the foundation for pioneering approaches to experimentation that have endured for more than a century.

Edison's creation of his own independent laboratories put him ahead of his rivals. Through his travels to Britain in the early 1870s, Edison was exposed to sophisticated European laboratory equipment unavailable in America. Inspired, Edison envisioned creating a laboratory of his own that would set an unprecedented standard of excellence. At Menlo Park, he surrounded himself with combinations of equipment that had never before existed in one laboratory—some of which he designed himself. Borrowing from the American machine shop tradition he'd become familiar with during his telegraph years, Edison transplanted steam-powered machine tools used in traditional shop invention to the environment of a sophisticated scientific research laboratory. To this foundation he added a world-class collection of chemicals and chemistry equipment of every description. Edison also included a woodworking bench with hand tools for developing prototypes, as well as steam-powered lathes and drill presses for metal work. Using this unique and cutting-edge assembly of equipment, Edison envisioned churning out an invention "every ten days, and 'a big thing' every six months." At Menlo Park, Edison realized his dream of creating an "invention factory."

Over the course of his career, Edison's dedication to equipping his laboratories with the highest quality machines and materials inspired the confidence of his workers, customers, and investors. His biographers noted that he was ". . . determined to have within his immediate reach the natural resources of the world." And, Dyer and Martin affectionately refer to him as ". . . the living embodiment of the spirit of the song 'I Want What I Want When I Want It.'"

Having set the stage for breakthrough innovations with his first state-of-the art facility, Edison taught his people to use his approach to experimentation. Like their boss, they learned to develop diverse hypotheses and to carefully examine their underlying assumptions. They were trained to execute experiments with meticulous care, and to thoroughly document each step. By rigorously training his staff in his experimental methods Edison made the process of innovation more systematic and reliable.

One of the most extraordinary examples of Edison's prowess in experimentation is his development of the first alkaline storage battery. In the late 1800s, most batteries were large and cumbersome. They generally contained lead—which made them heavy—as well as dangerous compounds like sulfuric acid that sometimes leaked. Leakage from these wet batteries often caused property damage and could burn the skin. Edison was determined to create a light and portable battery that contained no lead or liquid compounds. He designed experiments to test the broad range of hypotheses he had generated as a result of reading widely about alternative substances he could use to create a dry, portable storage battery. He then used the data from these experiments to generate a new series of hypotheses and a new series of experiments.

Edison's greatest challenge was finding a way to create perfect conductivity in the positive pole of the battery. His persistence in experimentation yielded an initial solution only after nearly a year of testing. He discovered that by using a unique sequence of chemical baths, he could create paper-thin flakes of pure nickel and iron oxide. He wrapped these around a metal core by devising a revolutionary technique to create thin "sandwiches" of nickel and iron oxide wafers compressed into sheets roughly one-eighth of an inch thick. His persistence in experimentation yielded an innovative product that set new standards of safety, effectiveness, and economy in portable power generation. The Edison storage battery proved to be a major success.

This circular advertisement positioned the Edison Storage Battery to the business market, offering the battery for use in trucks, manufacturing equipment, and other industrial applications.

Here is how one laboratory employee described "the hunt" for the storage battery:

> When asked how many experiments had been made on the Edison storage battery since the year 1900: "Goodness only knows! We used to number our experiments consecutively from 1 to 10,000, and when we got up to 10,000 we turned back to 1 and ran up to 10,000 again, and so on. We ran through several series—I don't know how many, and have lost track of them now, but it was not far from 50,000."

Edison described his inexhaustible determination to find an answer by saying, "When I have fully decided that a result is worth getting I go ahead of it and make trial after trial until it comes." Edison designed all his experiments to "surprise Nature into a betrayal of her secrets by asking her the same question a hundred different ways."

Edison's persistence in experimentation was also evident in his discovery of a commercially viable filament for the incandescent electric light. After painstaking initial research Edison focused on fibers taken from the bamboo family of plants. Bamboo's uniform cellular structure and ability to burn evenly when compacted led Edison to conclude that it was the optimal filament for commercial production. Edison had tested the properties of nearly one thousand species of bamboo plant by 1880, and identified those that were most promising. He began consulting his colleagues for the perfect person "to ransack the jungles of the Far East" in a quest for an adequate supply.

Edison ultimately selected William H. Moore of Rahway, New Jersey, as the right man for the job. Moore eventually found the ideal bamboo species growing in a farmer's grove near Kyoto, Japan. After testing thousands of specimens on location, Moore shipped the best samples back in bales to Menlo Park, where they were subjected to further comprehensive testing. Moore's hunt yielded the perfect fiber. As Edison later wrote to Moore, "Your trip to China and Japan on my account to hunt for bamboo and other fiber, was highly satisfactory . . . you found exactly what I required."

As Dyer and Martin describe it,

> It is doubtful whether, in the annals of scientific research and experiment, there is anything quite analogous to the story of this search and the various expeditions that went out from the Edison laboratory in 1880 and subsequent

years, to scour the earth for a material so apparently simple as a homogenous strip of bamboo, or other similar fiber.

Once the right kind of bamboo was located, carbonized filament samples had to be prepared. This meant kneading, mixing, and then rolling the material into filaments as fine as seven-thousandths of an inch in cross-section. One day, Charles Batchelor—one of Edison's top experimenters—had worked on kneading the lampblack used to carbonize the filament for what felt like an interminable period. He brought his sample to Edison and, as Dyer and Martin recount (describing Batchelor as the "assistant"), the following conversation took place:

"There's something wrong about this, for it crumbles even after manipulating it with my fingers."

"How long did you knead it?" said Edison.

"Oh! More than an hour," replied the assistant.

"Well, just keep on for a few hours more and it will come out all right."

Indeed, after a few more hours of kneading, Batchelor delivered to Edison exactly what he needed. Dyer and Martin report that the uncooperative mass of lampblack and tar had ". . . changed into a cohesive, stringy, homogenous putty." This sticky putty became the essential ingredient in carbonizing the spiral filaments that were then patented and used in the first commercially successful incandescent lights.

> There is always a way to do it better . . . find it.
> —Thomas Edison

"Persistence in experimentation" is the "sweat" that Edison cites in his classic aphorism: "Genius is one per cent inspiration and ninety-nine per cent perspiration." Edison frequently made this and other statements about the importance of hard work, commitment, and perseverance because he wanted to correct popular misconceptions about the process of innovation as "magic" or "wizardry." As Dyer and Martin explain it:

A popular idea of Edison that dies hard, pictures a breezy, slap-dash, energetic inventor arriving at new results by luck and intuition, making boastful assertions and then winning out by mere chance. . . . The real truth is that while gifted with unusual imagination, Edison's march to the goal of a new invention is positively humdrum and monotonous in its steady progress. . . . If, for instance, he were asked to find the most perfect pebble on the Atlantic shore of New Jersey, instead of hunting here, there, and everywhere for the desired object, we would no doubt find him patiently screening the entire beach, sifting out the most perfect stones and eventually, by gradual exclusion, reaching the long-sought-for pebble; and the mere fact that in this search years might be taken, would not lessen his enthusiasm to the slightest extent.

Edison's big goals, charismatic optimism, and thirst for knowledge all fueled his persistence in experimentation, and strengthened his solution-centered mindset. His approach to experimentation gave him a critical competitive advantage. By maintaining meticulous records of findings from his countless experiments, Edison created a formidable database of knowledge. This remarkable database, coupled with his reading, fueled Edison's extraordinary creativity in generating a broad range of hypotheses. The depth of his creativity was also evident in the amazingly innovative designs he developed for his experiments. The breadth of Edison's experimental approach is part of what enabled him to successfully undertake so many different forms of innovation. By training his people to execute experiments in a comprehensive, creative, and disciplined manner, he made a major leap toward the development of systematic innovation.

Creating Innovation Literacy: Experiment Persistently

> All life is an experiment. The more experiments you make, the better.
> —Ralph Waldo Emerson

Although Edison's primary experimental playground was his laboratory, he did not confine his experimentation to the lab. He conducted experiments in all kinds of places; in big cities and small communities, in the desert, parks and

Completely deaf in his left ear, Edison placed his face directly up against a working phonograph so he could feel the vibrations run through his facial bones. Edison used this same kinesthetic approach when working with telegraph equipment.

gardens—anywhere he needed to find an answer. Edison did not limit his idea of *where* experimentation—or innovation—could take place. His *solution-centered mindset* led him to learn everything he could from all his experiences.

Whatever the context for your experiments—the laboratory, office, your backyard, a classroom, a retail store, the Internet, or another venue of daily life—persistence, as Edison demonstrated, is an essential element for long-term success. Persistence in experimentation is key to achievement in every field, and a critical determining factor in the realization of *all* innovation. Curt Carlson and Bill Wilmot of SRI International—a leading research and development think tank in Menlo Park, California, responsible for breakthroughs including HDTV, robotic surgery, and the first computer mouse for Apple's Macintosh—are strong advocates of this Edisonian element. In their book *Innovation*, they emphasize that innovative solutions will only come to those who "iterate, iterate, iterate."

Presidents, poets, and philosophers praise persistence pervasively. Calvin Coolidge became president when Edison was seventy-six years old. He paid homage to the value of this crucial Edisonian quality: "Nothing in this world can take the place of persistence. Talent will not; nothing is more common than unsuccessful people with talent. Genius will not; unrewarded genius is almost a proverb."

Another Edison contemporary, Winston Churchill, expressed the essence of his life philosophy succinctly, "Never give up, never give up, never, ever, give up." And he added wryly, "Success is going from failure to failure without loss of enthusiasm."

Why do successful people from all walks of life emphasize the fundamental importance of persistence? One trait that successful people share is the tendency to study the lives of other successful people. Edison studied the lives of Paine, Robert Ingersoll, Faraday, Franklin, and Lincoln, among others. He discovered, of course, that they all overcame adversity, learned from their mistakes, and persisted tenaciously in their respective endeavors.

Great achievers understand intuitively that the human brain is the most profoundly powerful solution-finding mechanism in the known universe. And they recognize that persistence is the key to keeping that mechanism engaged. As we've seen, optimists get better results in life; and the main reason is simply because they are less likely to give up. As Dr. Martin Seligman emphasizes, pessimism is self-defeating because it "short-circuits persistence." Of course, it's much easier to be optimistic when everything is going your way, but the real key is, as Churchill reminds us, to maintain our enthusiasm in the face of seeming failure. **Resilience in the face of adversity is the greatest long-term predictor of success for individuals and organizations.** And, persistence in the process of experimentation, when desired or expected results are elusive, is the way that resilience is expressed.

> Persistence is to the character of man as carbon is to steel.
> *—Napoleon Hill*

ELEMENT 5: PURSUE RIGOROUS OBJECTIVITY

> We can all be outstanding scientists. . . . The key ingredient is to have the courage to face inconsistencies . . . This challenging of basic assumptions is essential to breakthroughs.
> —*Eli Goldratt*, The Goal

Edison's charismatic optimism about the achievement of his big goals was balanced with an almost Zenlike quality of detachment in regard to the results of individual experiments.

At an early age, Edison trained himself to view the outcomes of his experiments as "neutral" rather than negative or positive. He recognized that each experiment brought him one step closer to the answers he sought. By maintaining an objective viewpoint, Edison analyzed his findings without an agenda, applying the broadest possible applications of his results, thus yielding insights that would otherwise have been unavailable to him. The ability to look at one's self and one's circumstances objectively is a profoundly valuable tool for personal fulfillment and success. And, the ability to be a rigorously objective observer of your data is absolutely essential to the innovator.

Edison's persistence in experimentation was driven by his optimism. His passionate commitment to big goals inspired his insatiable quest for learning. And by continually striving to place his desire for truth above his need to be right, Edison was able to discover the broadest possible applications of his results.

Fascinated by all outcomes, Edison viewed each from a position of neutrality. The outcome of every experiment stood or collapsed on its own merits. His optimism about the long-term big picture was complemented with a disciplined approach to seeing things as they are.

Edison's willingness to maintain a neutral view allowed him to see unexpected patterns in his data, connecting findings from one experiment to outcomes in another—even trials conducted years apart for entirely different purposes. By documenting the results of each experiment, Edison's laboratories created vast databases of knowledge and intellectual property unrivaled by his competitors. Edison's objectivity and his openness in "connecting the dots" of his findings allowed him to discover patterns that accelerated his innovation process.

In one experiment conducted in England in 1873, Edison sent a single Morse code "dot" through miles of coiled undersea cable. The single dot generated a telegraph tape printout twenty-seven feet long! This unexpected and bizarre result fascinated Edison. In attempting to identify the source of the distortion, he designed a rheostat (a long tube designed to be filled with pressure-regulating substances) filled with soft carbon. But in Edison's laboratory experiments with the carbon rheostats, he noted they responded to the intense pressure of an undersea environment by varying with it rather than maintaining a static state. Another fascinating outcome! Carbon thus was not the answer to Edison's undersea signal distortion challenge.

Three years later, however, a few months after Bell's 1876 introduction of the telephone, Western Union asked Edison to improve the quality of Bell's poor telephone transmitter. Edison knew precisely where to begin: carbon. As a result of diligently tracking his experiments, including the bizarre twenty-seven-foot "dot," Edison was able to rapidly design the carbon button transmitter using carbon granules and a vibrating metal diaphragm. The carbon transmitter became an important improvement in American telecommunications, ultimately serving as the industry standard.

Edison made history with the carbon button transmitter and later the phonograph because he was able to discern patterns in the outcomes of results achieved in one category of work and connect them to findings in another. Because he was willing to be neutral and allow patterns in the data to emerge rather than forcing his own expectations upon them, Edison was able to accelerate the development of one industry—telecommunications—and pioneer another— entertainment.

Edison never tossed away any findings as useless. Every experimental outcome represented knowledge. Even if the findings seemed completely outlandish and without explanation, Edison respected the data, refusing to destroy it or set it aside. Instead, he chewed on the results even harder.

> At Whirlpool, named one of the world's top 100 most innovative companies in 2006 by *Business Week*, managers have learned not to discard ideas that didn't work "in their first go-round." Executives don't kill unsuccessful product concepts. Instead, they shelve them so that employees and managers can review them years later, when consumer tastes have changed and new conditions exist in the marketplace.

In several instances during his career, Edison discovered scientific anomalies while experimenting with technologies intended to advance an invention. The word "anomaly" is derived from the Greek, *an* meaning "not," and *homalus* meaning "regular." Thus, an "anomaly" is something that is irregular, an exception. In 1875–76 as Edison was experimenting with acoustical telegraph systems using vibrator magnets, he and his teams discovered something exceptional:

> They noticed a spark passing between the cores of the magnet and the lever [of the telegraph arm] that was similar to sparks they had frequently seen "in relays, in Stock printers when there were a little iron filings between the armature & core & more often in our new electric pen."

When this phenomenon had occurred previously, Edison and his staff had always attributed it to induction, "but when we noticed it on this vibrator it seemed so strong that it struck us forcibly there might be something more than induction."

In testing the phenomenon further, the laboratory team soon found that they could get the spark by touching the vibrator with a piece of iron and that the "larger the body of iron touched to the vibrator the larger the spark." In testing whether this was an induction effect, they discovered they could elicit a spark "from pipes anywhere in the room," and even draw sparks "by placing a piece of metal as much as three inches from the end of the lever." After discovering that it failed to register on a galvanometer and did not have any taste—signals that would indicate the phenomenon was an inductive force—Edison concluded that "the cause of the spark is a true unknown force."

Edison realized his findings were inconsistent with any outcomes he had ever logged. The inexplicable patterns in the outcome had shown that information transmitted through telegraph lines first became magnetic, then transformed into heat energy, then transformed back to electrical waves. By considering all his data with rigorous objectivity, and resisting the temptation to throw away these anomalous findings, Edison had unknowingly discovered high-frequency radio waves.

Anomalies, and other unexpected outcomes, represent important learning opportunities for the innovator. Not all results can be classified neatly. We want to stuff them into categories, and lump them into pre-existing classifications we already understand. The discipline of rigorous objectivity keeps the mind open to accept and explore unexpected solutions.

Creating Innovation Literacy: Pursue Rigorous Objectivity

Edison's objectivity allowed him to see patterns in his outcomes as well as accept anomalies—all leading him to make connections that led to his great innovations. He knew that it was essential to consider all the data he and his teams collected without prejudice. **The ability to consider data without prejudice is a distinguishing characteristic of the greatest minds.** This ability is a skill that can and must be cultivated by anyone interested in innovation.

But, the tendency to prejudge data, or slant the interpretation of it according to a predetermined agenda, is all too common in business, government, and academia. Profit data is sometimes skewed to meet Wall Street's expectations. Intelligence agency reports are occasionally tweaked to fit administration policies. Students can be graded based on their fealty to their professor's perspective rather than their own insights. When objective data is subjected to "spin," it is often motivated by the less-illumined realms of the psyche.

In the 1920s, managers at a Western Electric plant in Hawthorne, Illinois, conducted a study to determine the ideal levels of lighting to optimize productivity for the different parts of their factory. The factory workers found out that the study was taking place, and may have altered their productivity in response to the expectations of management. This response, known as the Hawthorne Effect—the notion that the expectations or actions of the experimenter may skew the outcomes of the experiment—became a cautionary principle in organizational research. Ultimately, social scientists cast doubt upon the original research, but those who did the work questioning the validity of the effect were suspected of having an agenda to disprove the Hawthorne Effect.

We now know with certainty that, at the level of quantum physics, all our knowledge is uncertain. Physicists confirm that on the most fundamental level, our perceptions and expectations do have a determinative effect on what we observe. And, in our daily lives we are often sent reeling from the onslaught of spin from politicians, advertisers, and interest groups. How, then, are we to pursue rigorous objectivity in an uncertain, relativistic, subjective, media-saturated milieu?

First, with humility; but also with the recognition that thinking with rigorous objectivity is a discipline that helps us get better results. We can train ourselves to be relatively much more objective. As psychologist David Perkins states, "Our thinking tends to be hazy, hasty, narrow or sprawling—casual terms for impulsive.

Just like anything else, thinking skills require upkeep. If they aren't nourished, they'll fade away."

As Perkins suggests, thinking is a skill. It can be taught and it requires upkeep. Edward de Bono is a pioneer in the field of teaching thinking as a skill. His book *Six Thinking Hats* offers an elegant and powerful approach to developing your ability to pursue rigorous objectivity and strengthen your *solution-centered mindset.*

In de Bono's approach, each "hat" represents a different type of thinking, or way of approaching a problem. Putting on a hat encourages each individual to apply a specific perspective that may vary significantly from his or her standard viewpoint. By practicing the hats technique, you will become increasingly aware of habitual impediments to objectivity and you will find it easier to move in and out of an objective frame of mind.

As de Bono emphasizes, effective thinking is often derailed by the tendency to confuse different modes. As Edison knew, our ability to pursue rigorous objectivity is hampered when we prejudge data in an overly positive or negative way, when we fail to assemble all our data in a comprehensive fashion before forming opinions about it, and especially when we allow our emotions, operating below our conscious awareness, to color our perceptions. The Six Hats approach is a simple way to distinguish these different modes and keep them pure. This tool is useful for both personal and group solution finding and decision making.

Here's a brief introduction to the hats:

White Hat: White is the color of neutrality and openness. The white hat represents a purely objective focus on data collection.

Red Hat: Red is the color of passion and emotion. Attempts at objectivity are often derailed because of unconscious emotional contamination. The red hat legitimizes the open expression of emotions, gut reactions, and intuitions.

Black Hat: Black represents the negative or devil's advocate position. Black hat thinking focuses on logical arguments highlighting the weakness of an idea or proposal. If you don't scrutinize all your ideas with rigorous black hat thinking, reality will do it for you.

Yellow Hat: Yellow is the color of the bright sun. It represents the positive, optimistic viewpoint. Yellow hat thinking focuses on the benefits and upside of an idea or proposal. It represents the "angels' advocate."

Green Hat: The color green represents growth and this hat is associated with idea generation. In green hat thinking, we suspend judgment and criticism and go for quantity of ideas.

Blue Hat: The color blue represents the clarity of the open sky. It is the hat worn by the facilitator of the thinking process. The blue hat guides the appropriate application of the other hats. If, for example, the group is looking at all the benefits of an idea (yellow hat thinking) and someone expresses their doubts or concerns (black hat thinking) the blue hat will intervene to keep the group on track.

Most of our decisions are ultimately made with the red hat. We choose a course of action because it feels right. As management pioneer Alfred P. Sloan noted, "The final act of business judgment is intuitive." The critical distinctions, however, are whether we objectively assess all of the available information *before* relying on feeling; and, how we separate intuitive judgment from prejudice and preconception. The six hats allow us to bring awareness to our thinking process, so we can do a better job of making these distinctions.

Dr. Jim West describes his application of this element for *Innovating Like Edison:*

"For me, because I'm an experimentalist, a lot of things happen in the laboratory where you think nature is behaving in a certain way. And then you suddenly find that, 'Wow, nature doesn't behave that way! It's doing something different.' Well, it may pull my thinking apart. I have to say, 'Is nature showing me what I think it's showing me? Or, am I doing something wrong? Or, am I approaching this problem wrong?' Certainly, it means you have to keep experimenting, you have to bring that experimentation out, to make sure that there are no errors on the part of the experimenter, or the instrumentation, or the materials. Once I'm pretty satisfied that I'm doing things right, then the next step is not to throw that away."

Edison's *solution-centered mindset* began with the formulation of grand objectives expressing his deepest passions. His unwavering focus on "the bright side of everything" allowed him to keep his mind open to solutions that others deemed impossible. In the process he inspired his associates, investors, and stakeholders to go far beyond their perceived limitations.

The young Leonardo da Vinci once noted, "The knowledge of all things is possible." Edison agreed, and he sought the knowledge of all things relentlessly. This quest led him to develop a new model of practical experimentation that he pursued with incomparable persistence. His passionate, optimistic pursuit of his goals, hunger for knowledge, and uncanny stamina in the process of experimentation were tempered with a remarkably disciplined objectivity. These elements combined to make his mind a supercharged magnet for solutions. A *solution-centered mindset* is the beginning of innovation literacy. It sets the stage for you to learn Edison's approach to idea generation and creativity that we call *kaleidoscopic thinking*. It will propel you and your organization to unprecedented levels creativity in service of innovation.

Chapter Four

COMPETENCY #2— KALEIDOSCOPIC THINKING

> [Edison has a] remarkable kaleidoscopic brain. He turns that head of his and these things come out as in a kaleidoscope, in various combinations, most of which are patentable.
> —*Western Union patent attorney Edward Dickerson*

Thomas Edison loved ideas. He loved to generate new concepts, and to combine familiar concepts in original ways. He expressed his delight in idea generation and exploration when he commented, "I would like to live about three hundred years. I think I have ideas enough to keep me busy that long."

As Dyer and Martin observed:

> Edison's inexhaustible resourcefulness and fertility of ideas have contributed largely to his great success, and have ever been a cause of amazement to those around him. Frequently, when it would seem to others that the extreme end of an apparently blind alley had been reached, and that it was impossible to proceed further, he has shown that there were several ways out of it.

Edison reveled in the process of defining and solving problems from diverse angles, and this made his mind especially fertile. In a diary entry on July 12, 1885, he celebrates the workings of his own "mental kaleidoscope," which he used to "obtain a new combination of ideas."

Kaleidoscopic thinking is our phrase for Edison's extraordinary approach to idea generation. This competency is an innovator's toolbox for outgunning the competition through original thought.

The kind of creative thinking Edison espoused is often suppressed in most schools today. Fortunately, it can be relearned in adulthood.

The Elements of *Kaleidoscopic Thinking* are:

6. Maintain a Notebook
7. Practice Ideaphoria
8. Discern Patterns
9. Express Ideas Visually
10. Explore the Roads Not Taken

ELEMENT 6: MAINTAIN A NOTEBOOK

What do Leonardo da Vinci, Isaac Newton, Pablo Picasso, Charles Darwin, Marie Curie, Albert Einstein, and Thomas Edison all have in common? They all kept notebooks.

In his teenage years as well as at Menlo Park and beyond, Edison recorded his thoughts, observations, and visualizations in notebooks. Like other great minds, Edison jotted down his thoughts freely. His notebooks contain fragments of ideas and plenty of pictures. This daily practice helped him sharpen his observations, develop new ideas, and make creative connections between diverse aspects of his research.

Edison loved nature and keenly observed his surroundings from the time he was a young boy. He continued recording his observations about the natural world throughout his life by jotting down his observations in notebooks. When Edison was twenty and living in Cincinnati, he was, according to Paul Israel, "in the habit of using small pocket notebooks to make 'notes and diagrams'" to record his thoughts about the world of nature, or noting the results of his experiments in chemistry and telegraphy. Edison and his laboratory staff generated more than 2,500 notebooks in his lifetime, most 200 to 250 pages in length, and ranging in size from 8.5"×6" to 9"×11". Edison always had a pocket notebook at the ready, and used these to augment those he maintained in his laboratory.

In addition to serving his creative process, Edison's notebooks also secured his intellectual property. In October 1870, Edison met Lemuel Serrell, a noted patent attorney. Serrell advised Edison to maintain careful records of his ideas, and that "such a record would be essential to defend his inventions in the patent office or

In 1928, Edison jots an entry in one of the more than 2,500 notebooks he generated with his staff. His distinctive handwriting makes Edison's notations easy to differentiate from those of his laboratory compatriots.

in the courts." After receiving this advice, Edison wrote in his notebook, that for "all new inventions I will here after [*sic*] keep a full record."

Edison took Serrell's counsel to heart and began dedicating notebooks for specific uses. In the summer of 1871, Edison created a series of four notebooks which would serve as official records "to be used in any contest or disputes regarding priority of ideas or inventions." He named the four notebooks as follows: 1) Gold & Stock Telegraph Co., with the first page reserved for "any ideas contained in this book which I do not see fit to give said G & S Telegraph"; 2) Record of Ideas day by day applicable and for the Dot and Dash system of fast Telegraph invented for Geo Harrington and D H Craig; 3) Dot and Dash and Automatic Printing Translating System, Invented for myself exclusively, and not for any small-brained capitalist; and 4) Ideas conceived, and experiments tried on miscellaneous Machines and things.

As Edison gained fluency in maintaining several notebooks for different purposes, he also began writing in many of them, "I do not wish to confine myself to any particular device." This statement not only served a legal purpose—allowing his ideas to be directed for multiple different kinds of patents—it expressed Edison's love for working on multiple projects simultaneously.

Edison's notebooks expressed the remarkable functioning of his kaleidoscopic mind. The notebooks made it much easier for him to make connections between his multiple projects and diverse areas of investigation. One of his most inspirational moments of connection came on his visit to the laboratory of inventor William Wallace in September 1878. In a flash, Edison instantly connected all he knew to that point about electricity and incandescence. He realized that an electrical current could indeed be subdivided and he saw how it could be done. As he phrased it, "I have struck a big bonanza." This bonanza of inspiration was, of course, predicated on the perspiration he had invested through his reading, experiments, and notebook work. And what did Edison do when the inspiration flashed in his mind? He immediately grabbed his notebook and began scribbling ideas.

Through Edison's use of notebooks, the tangible world of experimentation met the intangible world of imagination. The award-winning linguistics research of University of New Mexico professor Vera John-Steiner, author of *Notebooks of the Mind*, helps us appreciate this essential aspect of Edison's kaleidoscopic thinking. John-Steiner's research shows that notebooks can provide a crucial bridge between the raw thought of the "inner world" of our minds and the "outer world" of speech. John-Steiner helps us understand the value of

Edison's notebook drawing of a telegraph escapement mechanism, dated 1871.

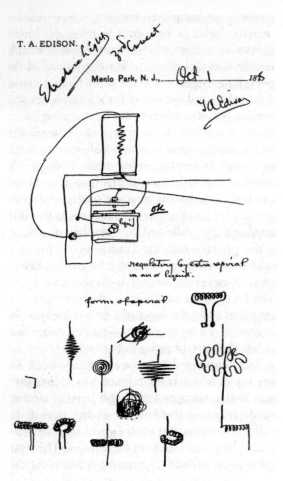

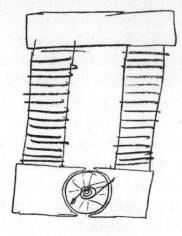

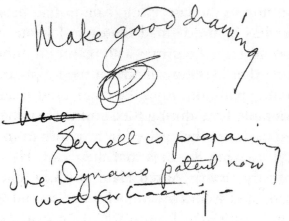

Notebook drawing for an electric light caveat
(a caveat is a patent modification), filed October 1,
1878, revealing Edison's love for looking at a
problem in multiple ways—in this case, the design
of a spiral filament.

This notebook drawing of a dynamo design for a
British patent filing reveals Edison's willingness to
abandon detail when necessary in favor of capturing
the overall idea.

recording ideas, as Edison did, in a fragmentary and incomplete fashion. She
explains that even though our internal thoughts are often fragmented, clumsy,
and imperfect, they are also highly symbolic. Raw ideas represent a "highly condensed language of thought where each word may stand for manifold ideas." In
terms that Edison would particularly enjoy, she points out that our inner speech
is like "telegraphic thought," where "a single word is so saturated with sense that
many words would be required to explain it in external speech." Use of this tele-

graphic style in keeping a notebook "makes it possible to gallop ahead, exploring new connections" without needing to "stop and explain specifics in precise, readable prose."

Take a look at the Edison notebook entries shown on pages 87 and 88. Here you can see Edison's "telegraphic thought" represented in both words and sketches. They are not neat and tidy. Words and pictures are often scrawled on the page, imperfectly formed. This is the "saturated sense" that John-Steiner describes, where each word and image is fuller in meaning than can be described in the moment of creation.

These notebook entries represent the raw beginnings of ideas that Edison later fleshed out into fuller form, ideas leading to innovations that changed the world.

Creating Innovation Literacy: Maintain a Notebook

Do you have to do any writing as part of your job? If you do, chances are that whatever you write will be read and evaluated by others. Most of the writing we do at work must have a beginning, middle, and end. It must demonstrate a clear logical flow or your message won't communicate its meaning to others. If you go on a business trip you will probably need to fill out an expense report when you return; even if you work at a very innovative organization, chances are that your boss is unlikely to suggest that you "Be creative!" or "Just make something up." Most of the writing that is done in the workplace—expense and other reports, proposals, plans, and memoranda—must be well reasoned, organized, and the numbers must add up properly. This kind of linear thinking is important to the everyday running of a business but it is not conducive to generating new ideas.

The beauty of maintaining a notebook as Edison did is that **you can express yourself freely in nonlinear fashion.** By recording your incomplete, fragmentary associative process you stimulate and inspire new highly saturated streams of thought. Begin by acquiring a blank-page journal or bound notebook. If you prefer, you can, of course, use your computer or PDA for the same purpose. You can dedicate different notebooks to specific subjects as Edison did and/or use one for all your observations. There are many ways to approach working with a notebook. Here are some tips to help you get the most out of your notebook.

- *Find the best time for you.* Some people like to devote a set time in the morning or evening to make their notebook entries while others prefer to use it at the moment they feel inspired. Experiment with the timing that works best for you.

- *Generate first, then organize.* A notebook invites you to freely express all your thoughts and feelings without having to hold anything back. Unlike business or academic writing, no one will criticize or judge it. Avoid allowing your own internalized critic to censor or edit your writing. If your notebook process leads you to generate an idea that you want to develop for practical use, you can then choose to critique it.

- *Use your notebook to record information.* As you record ideas, facts, stories, quotes, definitions of vocabulary words, jokes, and any other information that you find intriguing or inspiring, you'll discover that the recording process helps stimulate your production of new ideas and associations.

- *Doodle and draw!* In addition to words, the notebooks of Edison, Darwin, Leonardo, and many other great minds contain sketches, doodles, and drawings. Playing with images will stimulate your imagination.

- *Choose a theme.* Your notebook invites you to "free associate" and move from one topic to another. But sometimes a theme can help inspire and motivate you to greater depth. Julia Cameron and Mark Bryan, authors of *The Artist's Way at Work,* offer many practical work-related themes for notebook exploration, such as: How do I stay creative in a hostile and competitive environment? How can I remain creative despite criticism? How can I clarify and apply my strengths to my work? How can I handle an impossible workload?

- *Experiment with stream of consciousness.* Start writing in your notebook—or typing on your computer—and don't stop for at least ten minutes. Just keep the words flowing onto the page even if it seems that you are generating gibberish. This is a great way to burn through your habitual associations as you generate new connections.

Dr. John Wai, Director of Medicinal Chemistry at Merck & Co., Inc., comments on the value of keeping a notebook:

Writing down my ideas frees up my brain from remembering so many details, especially if the ideas are branching out in many directions. It's like a sketchbook of an artist, or a burst of beautiful harmonies or chords written down by a composer. To me, a notebook is a medium to nurture and capture my creative thinking, not an archive of my thoughts. The contents range from random thoughts, to detailed analyses, to how best to present certain things visually. There are also a lot of half-baked ideas and dead-end mental exercises!

Like Edison, Wai translates his nonlinear musings into practical innovation. He comments,

> For two years, our team was focusing on a series of promising leads for optimization as drug candidates. For the same two years, I thought of many other distinctly different structures that could offer significant advantages to what we were working on. As resources became available, I went through my notebook and worked on two that I perceived as better ones. Both worked!

ELEMENT 7: IDEAPHORIA

> To have a great idea, have a lot of them.
> —*Thomas Edison*

The word "idea" comes from the Greek *idein*, which means "to see." Euphoria is a state of delightful well-being. Ideaphoria is a neologism for the delightful well-being that accompanies the effortless flow of insights and ideas. We use the term to refer to Edison's approach to generating new ideas. He used three primary methods to generate ideas rapidly, and on command: word association; analogical thinking; and fantastical storytelling. These methods are relatively easy for adults to learn and apply, and all will generate tremendous results.

Edison began his three-stage ideaphoria process with raw association. Through using his notebooks, he created associative linkage between the ideas and experiences he already had with new ideas inspired by what saw in the world around him, or in the laboratory. He formed a habit of noting down his thoughts and regularly reviewing what he had written the previous day. He also regularly reviewed material from prior weeks and even years. His reviews inspired further torrents of possibilities and, as ideas flowed, he jotted them down in his notebooks. He didn't worry about whether he was right to begin thinking along a particular path because he knew that no matter where he entered the process, the results he sought would eventually emerge through chains of ideas all linked together. The order did not matter to him at this stage.

Edison's associative thinking "produced page after page of possible approaches to the problem" at hand. A member of his staff who'd worked with Edison for twenty years remarked: "Edison can think of more ways of doing a thing than any man I ever saw or heard of."

In one classic case, Edison asked one of his engineers to submit some sketches representing possible approaches to the creation of a new piece of specialized ore-milling machinery. The engineer generated three drawings that he promptly submitted to his boss. Edison wasn't satisfied, but the engineer protested that there was no other way to proceed. As Dyer and Martin recount the conversation, "Mr. Edison turned to him quickly and said: 'Do you mean to say that these drawings represent the only way to do this work?' To which he received the reply: 'I certainly do.'"

This exchange took place on a Saturday afternoon. After a day off, Edison stopped by the engineer's desk first thing on Monday morning and casually handed him a folder that contained *forty-eight different designs* for the new equipment. And, Edison's prolific idea generation wasn't just an academic exercise; one of his sketches formed the basis for the successful development of the new equipment.

Although it was rare for Edison to boast about his own genius, he admitted to being prolific with ideas. "I speak without exaggeration," he noted to a reporter, "when I say that I have constructed three thousand different theories in connection with the electric light, each one of them reasonable and apparently likely to be true."

Analogical Thinking

Analogical thinking is a way to generate insights by bringing together ideas that at first seem quite different from one another, but are later seen to be related in some way. From his early days in Port Huron and throughout his adult life, Edison reveled in literary analogy and metaphor. Edison's enthusiasm for classical literature, particularly Victor Hugo, led his telegrapher chums to give him the nickname "Hugo." Edison particularly admired Shakespeare's use of metaphor and analogy. Shakespeare's famous line, "Now is the winter of our discontent," appears in several of Edison's notebooks. In this metaphorical sentence, a season—winter—is likened to an emotion—discontent. The bringing together of these two disparate ideas yields an image that transcends simply saying, "It was cold and depressing outside."

Ah, Shakespeare. That's where you get the ideas!
   —*Thomas Edison*

Edison believed analogical thinking was fundamental to his invention process. In a 1915 interview, he stated that he considered "'a logical mind that sees analogies' to be an essential quality of an inventor . . ." When developing the incandescent light bulb, Edison realized that the flow of electricity through wires or other filament substances he was examining was like the flow of a message through telegraph equipment.

When developing a high-efficiency generator that more than doubled the output of existing generators, Edison "treated the magnetic lines of force in [the] generator as analogous to the internal currents of a battery and compared the flow of current in an armature with that in a battery."

Analogies also played a critical role in Edison's invention of the phonograph. Edison's deep knowledge of telegraphy allowed him to think of the telephone as a form of telegraph. Grappling with how to produce a written record of incoming telephone messages, Edison envisioned using a recorder similar to the embossing recorder-repeater he was then developing for Western Union. He began asking himself how the recording of telephone messages from a telegraph-like instrument could behave in ways similar to the recorder-repeater. He then drew an analogy between the operation of the recorder-repeater and the indentations created by mechanical embossing technologies he'd developed for the electric pen. All these analogies came together in Edison's "talking machine." And then he used analogy again in envisioning the creation of the motion picture camera, ". . . an instrument which does for the Eye what the phonograph does for the Ear."

Contemporary research into the nature of practical intelligence confirms that analogy is one of the mind's most powerful problem-solving tools. John Clement of the University of Massachusetts has conducted wide-ranging research on how higher-order thought operates. Much of it begins with analogical thinking. Clement discovered that a key element in the mind of an innovator is the ability "to generate analogies both within and across disciplinary boundaries."

Fantastical Storytelling

Edison's third ideaphoria technique was to write fantastical stories and "talk out loud" about speculative ideas that fascinated him. In 1890, Edison agreed to hold a series of interviews with George Parsons Lathrop, a well-established reporter who married the daughter of famous American author Nathaniel Hawthorne. Lathrop, who had previously served as editor of the *Atlantic Monthly*, held an abiding belief in the power of ideas, as demonstrated by his establishment of the American Copyright League to successfully secure international copyright laws. The purpose of Lathrop's interviews was to explore the great inventor's mind, his methods, and his musings about the future. These he ultimately compiled into a single work entitled, "Talks with Edison," which met with popular acclaim.

Among the extraordinary ideas they discussed were Edison's thoughts about atoms and molecules. At the time of Edison's interview with Lathrop, atomic theory was embryonic, Einstein was eleven years old, and Newton's laws reigned supreme. Remarkably, Edison's theories and ideas about controlling particles at small scales began to sound like a modern description of nanotechnology, or even genetic engineering. Lathrop describes their conversation:

> But in addition to being extremely practical in his thoughts and processes, Edison has a rich imagination of a creating sort, and moods of ideal dreaming . . . One day at dinner he suddenly spoke, as if out of a deep reverie, saying what a great thing it would be if a man could have all the component atoms of himself under complete control, detachable and adjustable at will. "For instance," he explained, "then I could say to one particular atom in me—call it atom No. 4320—'Go and be part of a rose for a while.' All the atoms could be sent off to become parts of different minerals, plants, and other substances. Then, if by just pressing a little push button they could be called together again, they would bring back their experiences while there were parts of those different substances, and I should have the benefit of their knowledge.

The favorable public reception to Lathrop's "Talks with Edison" led to a proposal for a science fiction novel on the future, to be entitled *Progress*. Edison was to draft notes about what he felt would be true about the world in AD 2226, and Lathrop would do the actual writing. Ultimately, Edison devoted himself to other priorities and the proposal never came to fruition, but he did write one hundred pages of notes for the book, thirty-three of which still survive, and can be viewed online.

Edison's "fantastical stories" and musings stimulated his imagination and led him to conceive of things that seemed impossible to others, like incandescent light, talking machines, and moving pictures.

Using the three approaches of ideaphoria—association, analogy, and fantasy—Edison suspended the world of everyday logic. He intuitively developed methods for following optimal brain pathways to generate a constant onslaught of ideas.

As neuroscientist Dr. Richard Restak states, "If you want your brain to function optimally, eliminate the tendency to deal with everything in strictly chronological terms . . . do away with the idea that the world must correspond to illusions of sequence and rational order." As Restak and other researchers emphasize, creativity isn't a mysterious gift; it is a natural human ability. Complexity experts Bill Welter and Jean Egmon, in their book *The Prepared Mind of a Leader*, point out that while creativity is complex, our "imagination is based on human capabilities and experiences each of us already has had and just need to practice reassembling in new ways." Edison's approach to ideaphoria provides a powerful guide for adults to practice "reassembling in new ways."

Creating Innovation Literacy: Ideaphoria

Please try this standard creativity test to benchmark your current level of ideaphoria.

"Alternative Uses" Exercise 1   In your notebook or on a piece of scrap paper, take two minutes and write down as many uses as you possibly can for a paperclip.

Take the total number of answers you wrote down and divide by two to calculate your score in terms of uses per minute.

The international average score is four uses per minute. A score of eight is excellent and a score of twelve or more correlates significantly with other genius-level measures of idea-generation ability.

When we give this test to groups of gifted children they invariably get genius-level scores, whereas most of our corporate clients generate average results. Why do the gifted children do so well? Because they immediately figure out that this is a test of *writing speed*. The test is purely quantitative. It asks you to think of "*as many uses as you possibly can.*" It doesn't ask you to think of "uses that you can defend before a board of directors, or senior management who will determine your pay based on the logical strength of your responses." But, most people in the organizational world interpret the instructions through this habitual lens.

Edison intuitively understood that free association was an important element in thinking outside the box. He knew that if you want to get a good idea you must first get a lot of ideas. Try this next exercise again now that you know this clue.

"Alternative Uses" Exercise 2   This time, in two minutes, write down as many uses as you possibly can for a brick. To think like Edison, focus on pure free association. Like the gifted children, treat this as a test of writing speed. Write down answers as fast as you can without analysis or criticism. Then, after you have generated a genius-level score, go back and use your imagination to explain your off-the-wall answers.

The low score that most corporate folks get initially is indicative of an organizational pandemic that interferes with innovation efforts at all levels. Chances are that you and your colleagues suffer from *premature organization*: the compulsion to organize one's ideas *before* generating them. Premature organization prevents conception and locks you into the box.

One of the simplest and most profound things you can learn from Edison is: **Generate first, then organize**. Edison's approach to analogical thinking and fantastical storytelling offers further support for your excursion into greater ideaphoria.

Analogical Thinking Exercise   The ability to see unexpected relationships and make unfamiliar connections was a delightful trademark of Edison's creativity. Linking things that seem to be unrelated is a wonderful way to awaken ideaphoria. Practice looking at things that, at first glance, seem unrelated and find different ways to link them. Or consider things that are obviously related and find connections between them that are not so obvious.

Experiment with drawing at least three links between the following things. There are no "right" answers in this exercise, only creative ones, so have fun.

- West Orange, New Jersey, and Shakespeare

- The light bulb and your job

- Radio waves and pasta

- The phonograph and religion

- Innovation and potato chips

- Portland cement and baseball

- Edison and the Internet

- Telegraphy and water

Dr. Robert Langer, a 2006 inductee into the National Inventors Hall of Fame, describes the role analogical thinking has played in his ability to solve complex problems. In overcoming the challenge of developing a polymer to facilitate a drug delivery mechanism that could release a medication over time rather than all at once, Langer asked, "What if we could develop a polymer that had a surface like soap—a surface that could erode harmlessly?" Guided by analogical thinking, he and his team succeeded in developing such a polymer, which could be delivered either through a pill, an injection, or even an embedded wafer. As he notes, "Just like soap, the polymer we developed dissolves layer by layer, which makes it very, very safe to use in the body."

Langer also recalls how analogical thinking spurred another breakthrough idea:

> One day I was just watching this TV show on PBS. I really wasn't even paying that much attention to it. I just kept seeing how they were making these chips for the computer industry, and I thought to myself, "Wouldn't that be a great way to deliver drugs! What if you could create a microchip that could be like a drug delivery device?"

With the help of his laboratory colleagues at MIT, Langer translated this analogy into a profound advance in the technology of drug delivery (US Patent 5,797,898), an advance that has benefited countless patients around the world.

Fantastical Storytelling Exercise: Image Streaming   Dr. Win Wenger has been researching genius for more than thirty years. Through his Project Renaissance, Dr. Wenger explores the most effective means for ordinary people to develop the knack of genius. One of his most intriguing discoveries is "image streaming." Image streaming is a deceptively simple way to energize your right hemisphere and emulate Edison's process of fantastical storytelling.

To begin, find a comfortable place to sit, and enjoy a few full, easy "sighing" exhalations to help you relax. Gently close your eyes, and then, simply describe aloud the stream of images that flows through your mind. To get the most from this simple but powerful practice you'll want to follow these important guidelines.

Describe the images aloud, ideally to another person or to a tape recorder. Silent description doesn't produce the desired Edisonian effect. Make your descriptions multisensory. If you see an image of a sandy beach, for example, be sure to

---

FANTASTICAL THINKING IN THE COSMIC MOLECULE

Thomas Edison described our solar system as a "cosmic molecule." His free-flowing imagination allowed him to generate ideas that were far ahead of his time. Edison used fantastical thinking to conceive of the following:

A transatlantic cable that could use the etheric force (high-speed radio waves, i.e., wireless technology) to transmit messages

Photography in total darkness

Electroplating in a vacuum

Producing electricity from coal

Artificial silk, leather, wood, and mother-of-pearl

Aerial navigation

High-speed trains

A single vaccination that could inoculate children against a variety of diseases

Antigravitation chambers (zero gravity)

Suspended animation

Space travel and interplanetary telegraphy

describe its texture, aroma, taste, and sound as well as its appearance. Of course, it may seem strange to describe the taste of a beach, but remember, this is an exercise in thinking like one of the most imaginative people who has ever lived. Descriptions in the present tense are more effective in eliciting vivid imagination, so express your flow of images as through they are happening in the now.

You can do image streaming without a theme as a free-form, spontaneous adventure in ideaphoria. Image streams usually gather their own momentum and express themes without your conscious instruction. And you can also use this technique to ask a specific question or explore a particular theme, as Edison did when he sent one of his atoms out to "become part of a rose." Dr. Wenger has used the method to develop numerous practical inventions and educational innovations.

ELEMENT 8: DISCERN PATTERNS

Edison cultivated an awareness of patterns in the world around him beginning in childhood. He came to believe that nature expressed itself in precise, mathematical patterns. Edison's belief in the omnipresence of these patterns gave him unshakeable confidence that he could ultimately decipher nature's codes. Edison revered the patterns in nature as expressions of God's handiwork. He was astonished and endlessly fascinated by the beauty and infinite connectedness he perceived in the natural world.

He cultivated his ability to discern patterns, and all of his technological discoveries rest in one form or another on this skill. Many of Edison's world-changing inventions were created through the recognition of patterns of connection between seemingly unrelated technologies. His passion to discover patterns helped him to understand trends, find gaps in the marketplace, and determine how technology could be applied to solve problems in new ways.

Edison began exercising his ability to discern patterns by filling in missing lines of press copy during his years as a telegrapher. As a novice telegrapher, he discovered that glitches in transmission were common. He often had to make up several sentences based on the overall pattern of the message, so that the messages would be complete. Edison quickly became remarkably adept at this, so that he could "write down what was coming and imagine what wasn't coming."

As an adult, Edison expanded his internal pattern database through his voracious reading, persistent experimentation, and by filling his laboratories with specimens from nature including vast collections of ores, chemical powders, bark, plant fibers, clays, exotic metals—thousands of compounds of every description. Although he believed that the ability to perceive patterns was available to anyone, he expressed dismay that more people did not discipline their minds to do it. He noted, "It is astonishing what an effort it seems to be for many people to put their brains definitively and systematically to work."

In his book *Blink*, journalist Malcolm Gladwell describes the ability of an expert art historian to rapidly determine the authenticity of an expensive Egyptian statue purchased years ago by the Getty Museum. The art historian studies the statue for a few moments, and confidently proclaims it to be a fake. Gladwell describes how this kind of instant—yet accurate—assessment happens in "the blink of an eye," and expresses an "intuitive knowing." Upon reflection, experts who make such summary judgments can usually find the basis for their conclusions. In the case of the art historian, it was the design of the fingernails that was not consistent with the pattern of the genuine article. His keen observation was an expression of pattern recognition.

Edison used pattern recognition in a similar way. He could look at reams of laboratory reports, or the summaries of experimental findings prepared by his employees, and detect erroneous data immediately. His mind seemed to process patterns in the data instantaneously. As one of Edison's master mechanics noted:

> Edison would examine the tabulated test sheets. He ran over every item of the tabulation rapidly, and, apparently without any calculation whatever, would check off errors as fast as he came to them, saying: 'You have made a mistake; try this one over.' In almost every case the second test proved he was right.

Edison's ability to detect errors quickly was manifest in his early experiments with wiring systems for the incandescent light. Edison began using copper for lead-in wires because of its exceptional conductivity and lower cost versus platinum. However, when he realized that several lighting experiments he'd conducted in exactly the same way had yielded different results, he seized upon the notion that the copper wire he was using must have had "dead spots." Immediately, Edison began cutting the wire into small segments, noting darker patches where he hypothesized that impurities had contaminated the copper during manufacturing. After chemical analysis, both Edison and the manufacturer realized he was correct. Edison saw in a flash that the pattern of results he obtained was inconsistent not because of his experimental method, but because of the quality of the materials. This insight was critical to the success of his lighting system, and initiated dramatic improvements in the manufacture of copper wire in the United States.

In 1879, after achieving his landmark 14.5-hour burn for the first filament, Edison began to project the trajectory of burn rates he felt were achievable with even better-quality materials. His ability to discern patterns in his data, then use these patterns to make projections, allowed him to achieve successively higher threshold levels of lamp life. His experiments also led Edison to begin calculating what kind of filament supply would be required for various levels of consumer demand, determining the economic viability of his work with incandescence at various consumer usage rates. His forecasts, like his acumen in detecting errors, proved to be remarkably accurate.

Edison's awareness of patterns also allowed him to see how the component parts of solutions he had generated in one area of experimentation could apply to applications in another completely different area. For example, to make the phonograph work, Edison had to find a way to transform sound waves into kinetic energy so the waves could be stored and replayed. He began experimenting to see how sound waves could be transferred to surfaces such as foil, wax, and other substrates without being distorted. Years of observing how incoming Morse code messages were tapped out by the chiseled stylus on most telegraph equipment led Edison, in another moment of inspiration, to see how this same chiseled stylus might be used to transfer sound waves onto cylindrical records. He applied the principles of a stylus to constructing phonograph needles that could activate the sound patterns etched into the grooves of his records, without damaging the grooves themselves. These remarkable achievements came because Edison discerned a pattern in the way indentations are made in both recording telegraph messages and in storing sound waves kinetically.

As he collected data, Edison would regularly ask, "Have I seen this anywhere else before? Is there a pattern here?" Edison's wide-ranging reading and experimentation enabled him to create a broad context for his observations. He thus could often discern patterns where others could not. He was open to seeing patterns emerge in all areas of his work.

As Malcolm Gladwell observed, the "aha!" we sometimes feel when seeing a pattern emerge is because a new connection has been established—one that is surprising and novel. These connections do not typically arrive through direct conscious effort, as with analogy, where relationships can be intentionally paired.

Dr. Richard Restak emphasizes that the ability to discern patterns requires an ability to shift context fluidly, and make new connections in a way that allows harmonious, fluid movement between the left and right hemispheres of the cerebral cortex. This is how we can "see the forest *and* the trees." In zooming out to assess macro patterns of the forest we call upon the facilities of the right brain. In zooming in to assess component parts of the trees, we call upon the facilities of the left brain. Edison's achievements in discerning patterns were a reflection of his ability to mobilize and coordinate these two modes. This whole-brain thinking skill is an essential element of innovation literacy.

Creating Innovation Literacy: Discern Patterns

What is one of the simplest and most powerful tools you can learn to improve your ability to discern patterns and promote the harmonious functioning of the hemispheres of your brain? Mind Mapping®!

Edison's ability to discern patterns began with his fascination with the natural world. If you contemplate the structure of patterns in nature as he did, you will see that a tree or a plant, for example, are networks of life, expanding in all directions from a trunk or stem. However, the most amazing network of all is right inside your skull. Each of your billions of neurons (brain cells) branches out from a center, called the nucleus. Each branch, or dendrite (a word that comes from the Greek word meaning "tree"), is covered with little nodes called dendritic spines. A gap between nodes is called a synapse. Our thinking is a function of a vast network of synaptic patterns.

A mind map is a graphic expression of these natural patterns of the brain. **Mind mapping is a simple, easy method for helping you discern patterns and make creative connections.** It was originated by British brain researcher Tony Buzan, who was inspired by his study of patterns in nature, recent brain research, and the notebooks of great minds like Leonardo da Vinci and Thomas Edison.

All you need to begin is a few different colored pens and a large sheet of blank paper.

Here's how to do it:

- *Begin your mind map with a symbol or picture (representing your topic) at the center of your page.* Starting at the center opens your mind to a full 360 degrees of association. Pictures and symbols are much easier to remember than words and enhance your ability to think creatively about your subject.

- *Use key words.* These are the information-rich nuggets of recall and creative association.

- *Connect the key words with lines radiating from your central image.* The branches will show connections clearly.

- *Print your key words.* Printing is often much easier to read and remember than writing.

- *Print one key word per line.* This can enhance the precision of your thought.

- *Print your key words on the lines, and make the length of the word the same as the line it is on.* This maximizes clarity of association and encourages economy of space.

- *Use colors, pictures, dimension, and codes for greater association and emphasis.* This will strengthen your memory and inspire your creativity.

Mind mapping makes it easy to have all your ideas for a topic on one piece of paper arranged in a way that encourages you to see relationships between them and discern patterns.

A senior research associate for a leading chemical company describes how he used mind mapping to discern patterns. He was attempting to "integrate a large amount of apparently unrelated data on a pulp bleaching process." After putting all the data on a mind map he commented, "As I began to make connections between the various elements of the process, I could identify and define an invention for which a patent is now pending . . . The mind mapping process took less than an hour and clearly was the key to defining and refining the new invention."

There are also many programs available for mind mapping on your computer. These programs can be very useful, especially for sharing mind maps with others via the Internet. We strongly recommend, however, that you learn and practice mind mapping "by hand" first.

Professor James Clawson of the University of Virginia's Darden Graduate School of Business commented in a recent conversation with the authors on the importance of discerning patterns in the management of innovation:

The ability to discern patterns out of oceans of disparate data has become a central skill of effective management. This ability, essentially *inductive logic*, asks that one become skilled at seeing the raw data in all its volume and chaotic naturalness—and infer from that data, as Edison did, the patterns that lie underneath. This skill is the essence of the scientific method. Unfortunately, much of our educational system does just the opposite . . . College and graduate school (MBA) students are too often told what the theories and/or principles are and then told to apply them to their attempts to manage. Many strategically challenged managers and executives want to apply deductive principles long after they're proven unproductive—because that's how they've been trained. The ability to discern patterns remains relatively rare but it's increasingly critical, especially for anyone who hopes to innovate.

Dr. John Wai, Director of Medicinal Chemistry at Merck & Co., Inc., describes the importance of discerning patterns in his team's efforts to develop innovative pharmaceuticals:

The process of pattern-seeking is very visual with drug design. I use computer modeling as well as plastic molecular models that I can touch and continually manipulate. One day, while looking for patterns in one particular class of inhibitors I'd been working on, it struck me that there was a unique element of symmetry in the molecules after dissecting them to the minimum. I then reconnected the crucial atoms back in an alternative way—using chemical bond patterns—and the structure revealed its secrets! From there my team was able to move forward with recommended testing for a unique new drug.

Discerning patterns gives us the creative edge we need to succeed. Awareness of patterns stimulates everyone on our team to think deeper. It urges us to expand and diversify the leads we pursue. It is analogous to the evolution from reptile to bird—only faster.

ELEMENT 9: EXPRESS IDEAS VISUALLY

> Some dreamed of a new alphabet, a new language of symbols through which they could formulate and exchange their new intellectual experiences.
>
> —*Hermann Hesse*, The Glass Bead Game

When you gaze into a kaleidoscope, you see all kinds of multicolored patterns. As you turn it the patterns shift, creating visual delight. Edison delighted in the process of exploring ideas visually in his kaleidoscopic mind. He translated his internal visions into an endless series of drawings. He developed many of his most creative ideas through the process of visual representation, a process that also included the creation of three-dimensional models and prototypes. Visual metaphor helped Edison look at his ideas from many angles, building from the known toward the unknown.

Edison always loved taking machines apart and then reassembling them. He also enjoyed scavenging for spare parts and then experimenting with different ways to combine them to create something new. As he explored how things worked he naturally wanted to discover how they could work better, and he began to use drawings to facilitate his understanding. Edison's visual explorations are extraordinary not only for their breadth and number, but because many are rendered from the perspective of an observer seeing the interior of an object from its exterior. Edison's visual "X-ray" imaginings allowed him to "see" how a yet-unperfected machine might function. His drawings capture the essence of how a machine might be able to work. And this is precisely why his mechanics could so successfully develop prototypes based on his sketches.

Edison used visual representations of ideas not only to build three-dimensional prototypes of his inventions, but to expand his thinking about what was possible. Through use of visual metaphor, Edison built out from what he did know toward what he did not know. In this process, he used images of concrete objects whose properties he understood and combined them with representations of concepts he did not understand. By imagining and sketching how different materials might function together in different circumstances—even wildly outlandish ones—he discovered new ideas.

Visual metaphors advanced Edison's thinking about the telephone transmitter, the electric light, the acoustic telegraph, and many other breakthroughs. On this

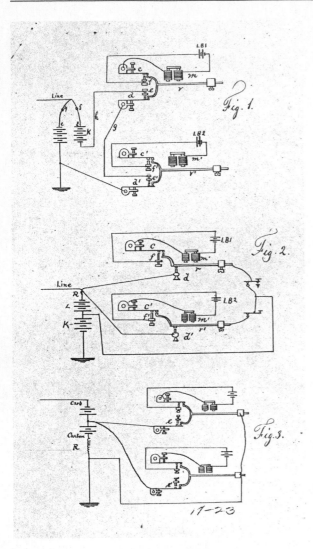

A detailed drawing used for filing Edison's Multiplex Telegraph Caveat #77, dated January 26, 1876.

and the following pages are three of Edison's drawings completed in the mid- and late 1870s during his investigation of the most extraordinary challenge of his career: subdividing an electrical current. The images he developed helped him explore the questions, "What could the circuit look like?" "How can I create a circuit architecture that works smoothly?" "What are the components that will enable the circuit to operate efficiently?" "How do all these electrical components relate to each other?" "How can I conceive of an electrical circuit—which I don't yet understand—to be like a telegraph circuit—which I do understand?"

The drawing on the left was completed on January 26, 1876, illustrating a Multiplex telegraph circuit. The Multiplex was an actual patented design of Edison's that allowed multiple telegraph machines to operate on the same line. Here Edison combined his knowledge of telegraphy and electromechanics in developing his visual representation.

Bearing this first image in mind, look at the drawing on page 107, completed more than two and a half years later on September 13, 1878. This drawing shows an electrical circuit that could split a single electrical current into discrete packets, powering individual lights. Edison again relies here on his knowledge of telegraphy to create a visual metaphor for his evolving ideas. He borrows the wiring techniques and structural architecture of a telegraph circuit, but employs principles of distribution within each individual circuit specifically related to electricity. Edison thus visually "experiments" with seven different structural combinations, each drawing upon his deep knowledge of telegraph mechanics, telegraph circuit regulators, and telegraph relay wiring.

The visual techniques he employed in this drawing advanced his thinking about electrical currents and filaments.

Roughly five weeks after completing this second drawing, Edison had learned more about the filament through additional laboratory experiments. The third notebook drawing shown on page 108, entered on October 25, 1878, reveals additional detail, particularly in the spiral designs of the filaments (also called "burners"). Although Edison's had not yet identified the optimal shape and size for the filament—hence the continued use of the telegraph metaphor—he had found that spiral filaments effectively dispersed heat, thereby increasing burn-life.

Together, these three drawings illustrate how Edison used visual metaphors to evolve his thinking from areas that he knew well toward areas that he did not.

Edison also used drawing as a way to play with evolving concepts and designs. He found drawing to be a practical way to "unstick" his mind. Edison's use of visualization in his mind and drawings on paper is consistent with decades of research demonstrating that vivid visual representation is one of the best ways to enhance memory and creative thinking.

Visual metaphors played a key role in breakthroughs achieved by Leonardo, Newton, Darwin, Einstein, James Watson and Francis Crick,

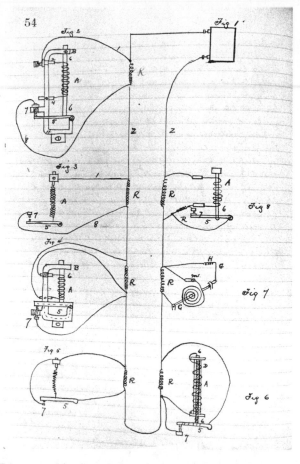

Edison used his knowledge of telegraphy to help him design circuits for electric lighting. Electric Light Draft Caveat #1, dated September 13, 1878, shows Edison's attempts to use the concept of a telegraph "regulator" used to break circuits when they become overloaded, as a guide for diverting electric current before overheating an incandescing filament.

and many others. Scientific historian Howard Gruber has conducted extensive research on the role of visual images in the creative process of great scientists. He states, "The scientist needs them in order to comprehend what is known and to guide the search for what is not yet known."

Edison's gift for visually exploring ideas on paper, also offered him an effective

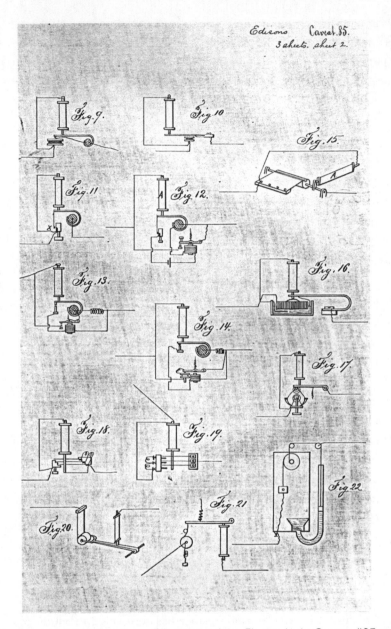

Six weeks and several iterations later, Edison completes Electric Light Caveat #85 on October 25, 1878, showing his incorporation of a spiral design for the filament to more rapidly diffuse heat from its incandescing surface.

way to communicate with his teams. Edison's staff had varying levels of formal education. Some had a technical education, some a master's degree, and some almost no formal education whatsoever. By using visual images, Edison transcended any communication barriers posed by written language, creating common ground for all. Drawings became the code of Edison's laboratory, the unspoken language everyone understood.

Edison shared his drawings with the laboratory staff to show them the entire scope of an idea on one page. Regardless of how he delegated the work for each invention, every team member could see what the entire concept looked like right from the start, understanding how his individual part fit into the whole. Edison used drawings to create a shared vision of innovation.

Creating Innovation Literacy: Express Ideas Visually

> The eye of the master will do more work than both his hands.
> —Benjamin Franklin

The best thing you can do to develop your ability to express ideas visually is to learn how to draw or paint. While not everyone can be a Leonardo, Picasso, or Georgia O'Keeffe, the good news is that **it is surprisingly easy to learn to *draw well enough* to facilitate the process of *kaleidoscopic thinking*.** Indeed, the main impediment to experiencing the benefits of expressing your ideas visually is the belief that drawing is difficult and something that only artists can do.

Children's drawings are remarkably artistic, expressive, and alive because they haven't yet developed the habit of criticizing and analyzing every mark they make on a page. You can experience a rebirth of your childlike enjoyment of drawing by learning to temporarily suspend your adult habit of self-criticism. If you allow yourself to play with images, doodles, color, and form in the manner of a child, you'll be surprised and delighted by the process and the result.

The most successful approaches to helping adults learn how to draw are based on strategies for accessing the same innocent artistic awareness you had when you were a child. Dr. Betty Edwards, author of *Drawing on the Right Side of the Brain*, has taught millions of people around the world how to draw. She guides her students to let go of their reliance on the verbal, critical, analytic "left-hemisphere dominant" mode. This mode is useful for criticizing art but not for making it. Instead, Edwards invites students to use the nonverbal, intuitive, "big picture" right hemisphere. Her simple, elegant methods make it easy to begin to see and then draw in a natural manner.

In his classic book *The Natural Way to Draw*, Kimon Nicolaides explains that the way to learn to draw is "perfectly natural." He emphasizes that, "It has nothing to do with artifice or technique. It has nothing to do with aesthetics or conception. It has only to do with the act of correct observation, and by that I mean a physical contact with all sorts of objects through all the senses." In other words, the ability to draw is largely a function of focused sensory awareness. As Frederick Franck, author of *The Zen of Seeing*, and many others have emphasized, drawing emerges naturally when we give deeper attention to the world around us. As Leonardo da Vinci proclaimed, all drawing springs from "knowing how to see."

Earlier you were introduced to mind mapping as a tool for discerning patterns. Mind mapping is also a marvelous way to express your ideas visually. As you experiment with learning to draw you can include more drawn images in your mind maps, and gain even more benefit from this *kaleidoscopic thinking* tool.

If you are reluctant to learn to draw and haven't yet incorporated the skill of mind mapping, you can, nevertheless, boost your ability to think kaleidoscopically and benefit from the visual expression of your ideas by engaging the services of a graphic facilitator. Graphic facilitators are more than just illustrators. Now hired by Apple, Intel, Procter & Gamble, among other Fortune 500 companies, they are skilled in using a variety of visual tools to help you develop a vivid, shared vision in strategy, product development, problem-solving sessions, and many other situations.

Dr. Jim West describes the importance of expressing ideas visually in his work:

"I'm very dyslexic, so I can't read something and learn only from the reading. If I look at a bunch of equations, the equations as they stand don't mean very much to me. It's how they graph out. Then I can understand what they're telling me in terms of the parameters that are involved. When I see something graphically, then I begin to understand it, but not until I begin to see it that way do I really fully understand it."

ELEMENT 10: EXPLORE THE ROADS NOT TAKEN

In the republic of mediocrity, genius is dangerous.
     —*Robert Ingersoll*

Innovation demands the ability to think independently and act courageously. Throughout his career, Edison had the courage to embrace views that were counter to prevailing belief. As Niccolò Machiavelli, one of history's great political innovators, wrote:

> It must be remembered that there is nothing more difficult to plan, nothing more doubtful of success, nor more dangerous to manage than the creation of a new system. For the initiator has the enmity of all who would profit by the preservation of the old institutions and merely lukewarm defenders in those who would gain by the new ones. The hesitation of the latter arises . . . in part from the general skepticism of mankind which does not really believe in an innovation until experience proves its value.

Edison often stood in opposition to "those who would profit by preservation of the old institutions." This element—*explore the roads not taken*—energizes the fruits of all the other elements of *kaleidoscopic thinking*.

Edison's challenge to conventional scientific thinking about incandescence provides a perfect example of his willingness to explore the roads not taken. Historians of science generally agree that the phonograph is the supreme reflection of Edison's originality; the quadruplex telegraph is the greatest expression of his technical finesse; the alkaline storage battery his most compelling example of persistence; and the kinetoscope motion picture system his most ingenious combination of technologies. But the incandescent electric lighting system is by consensus the most revolutionary expression of Edison's inventive genius. It was born from his

Edison became a living symbol of progress through technology. He poses here with an improved phonograph that he and his teams developed in response to intensifying competition.

courage to defy convention and to embrace a staunchly contrarian point of view.

In the 1870s, Edison saw that existing methods of providing light and power were messy and unsafe. Lanterns with gas, oil, grease, or other substances often led to nasty spills and fires. Edison envisioned a form of lighting that was easy and safe, as well as affordable. But "those who would profit by preservation of the old institutions"—namely the firms selling oil, grease, and gas—took every opportunity to oppose and even ridicule his efforts. Edison was undeterred. As he stated, "I shall make the electric light so cheap that only the rich will be able to burn candles."

In addition to opposition from those with vested interests, Edison faced the scorn and ridicule of luminaries in the world of science. In 1878, finding the means to heat any substance to a white-hot glow without destroying it seemed impossible. As Sir W. H. Preece, the chief engineer of Britain's post office, commented, Edison's effort would be "an absolute ignis fatuus." "Ignis fatuus" is a medieval Latin term for a light appearing over swampy ground at night caused by spontaneous combustion of substances such as methane gas, but it also carries the implication of something illusory and even foolish. Another distinguished expert proclaimed, "Much nonsense has been talked in relation to this subject. Some inventors claimed the power to 'infinitely divide' the electric current, now knowing or forgetting that such a statement is incompatible with the well-proven law of conservation of energy." And, Britain's prestigious Royal Institution concluded that practical incandescence was "utterly impossible."

On November 4, 1879, just over one year after the Royal Institution's pronouncement, Edison applied for his basic patent on the incandescent electric lamp.

Edison often embraced ideas that were outside the mainstream, including the works of Robert G. Ingersoll (1833–1899), a charismatic American orator and champion of "free thinking." Ingersoll inspired many original thinkers of the age, including Mark Twain, Frederick Douglass, and Oscar Wilde. When Wilde toured the United States, he attended several of Ingersoll's sold-out speeches and pronounced him to be "the most intelligent man in America."

Ingersoll reveled in irreverence. He said, "I'd rather smoke one cigar than hear two sermons." He preferred science to faith: "Any doctrine that will not bear investigation is not a fit tenant for the mind of an honest man," and, "In nature there are neither rewards nor punishments. There are consequences."

Ingersoll, Faraday, Paine, Lincoln and, of course, his father Samuel Edison all served as role models for Edison. They inspired him to challenge conventional thinking and to achieve the "utterly impossible." As stated by complexity experts Welter and Egmon, challenging convention at the level that Edison did,

> requires developing the guts and the courage to stand on your own convictions, and step out of the mainstream when necessary. Taking a disruptive path requires commitment to human values as well as a desire to fundamentally change/improve a given situation or thing. It demands the utmost in self-awareness, testing the depths of one's strengths and weaknesses.

Creating Innovation Literacy: Explore the Roads Not Taken

Innovation demands both cognitive and emotional intelligence. Thomas Edison is a paragon of the cognitive freedom and emotional fortitude required to innovate. As he observed, most people are unwilling to discipline themselves to think through new ideas. He commented, "There is no expedient to which a man will not go to avoid the labor of thinking." Edison's contemporary, the Irish dramatist George Bernard Shaw, makes the same point in amusing fashion, "Few people think more than two or three times a year; I have made an international reputation for myself by thinking once or twice a week."

The willingness to consider new ideas intellectually must be supported by the courage to champion them into reality. Few know more about championing practical innovations than Curtis Carlson, CEO of SRI International. In their book *Innovation*, Carlson and his collaborator Bill Wilmot emphasize: "There must be a champion who proactively identifies with the customer and who addresses the funding, bureaucratic, political, human, and technical challenges that every innovation faces." They add, "No champion, no project, no exception."

Dr. Donald Keck was inducted into the National Inventors Hall of Fame in 1993 for his co-invention of the first optical fiber. He comments on championing innovation in a big company: "One of the things that you find out early on working for a large corporation is that nobody has more interest in getting your invention through the pipeline than you do." Keck's ability to heed a different drummer was, as he explains, a critical element in this breakthrough. "Dr. Bob Maurer hired me into Corning right out of graduate school. Bob instilled in us the notion of being contrarian. As Dr. Peter Schultz—one of my other collaborators—and I thought back over the years, we said, 'You know, we took the road less traveled.'"

Keck adds,

We later found out that Bell Labs had a group of 20 or 30 people trying to invent the same fiber we were. British Telecom had a similar effort. I'm sure that Nippon Telephone and Telegraph in Japan had a major effort as well. But all of those labs pursued what I now term "the engineering approach." They took the very best that anybody knew at the time, and simply tried to improve it. But all our experiments told us we wouldn't get very far at all if we followed that line of thinking. We took a revolutionary path instead. There were actually three disruptive innovations that went into creating optical fiber. It turns out we took a contrarian direction on all the key pathways we pursued.

How do you cultivate the intellectual and emotional strength, like Donald Keck and his collaborators, to explore roads not taken? Do what Edison did and immerse yourself in the study of great, independent thinkers and innovators. We believe Edison is an extraordinary role model, but there are many others, including those who inspired Edison himself. Make a list of your most inspiring champions and learn as much as you can about them. As Mark Twain wrote, "Really great people make you feel that you too can become great."

> Do not go where the path may lead, go instead where there is no path and leave a trail.
> —*Ralph Waldo Emerson*

Edison's kaleidoscopic mind brought forward revolutionary ideas that changed the way we live. In bringing the world electric light, Edison bucked conventional wisdom. But he also had to develop an entire array of new equipment to drive electricity from a central power station out to homes miles away. He had to find investors to subsidize development of his work, price the equipment properly, protect his ideas, and not drive himself to an early grave in the process. Edison's ability to manage more than forty projects simultaneously at the height of developing his electrical power system stands as testimony not only to his exceptional *kaleidoscopic thinking* abilities, but his capacity for managing complexity—a key skill covered in Competency #3: Full-spectrum Engagement.

Chapter Five

COMPETENCY #3— FULL-SPECTRUM ENGAGEMENT

To *Innovate Like Edison* you've got to be tough *and* supple at the same time. Edison worked with incredible intensity, yet he could shift into a playful mode in an instant. He understood the importance of finding the balance between concentration and relaxation, a balance that is a profoundly liberating for your genius power. Edison learned to be aggressive in defending his intellectual property, yet he was open and generous with his ideas. He also knew an essential approach for unleashing his full potential for innovation, an approach that has become even more important today: how to discover simplicity and clarity in the midst of ambiguity and complexity.

As journalist George Parsons Lathrop noted:

> He has in a degree which is literally startling, the power of self-concentration. With him no time is wasted on formalities and conventions, and not an instant is lost in passing from one mood or subject to another. The transition, moreover, is made with the whole momentum of his mind. . . . He does everything with the least amount of friction.

Undertaking everything "with the least amount of friction" and with "the whole momentum of his mind" enabled Edison to operate in an almost constant state of "flow." Edison's ability to "flow" meant that he was able to engage fully along a remarkable spectrum of behaviors. We call this third competency *full-spectrum engagement.*

The Elements of Full-spectrum Engagement can be thought of as sets of twin forces that must be balanced:

11. Intensity and Relaxation
12. Seriousness and Playfulness
13. Sharing and Protecting
14. Complexity and Simplicity
15. Solitude and Team

ELEMENT 11: INTENSITY AND RELAXATION

Edison's desire to work "with the least amount of friction" drove the unique culture he established in his workplace. Twelve- to eighteen-hour workdays were common, particularly in the years following Edison's development of the incandescent light. Progress in the laboratory was measured less in terms of absolute hours expended as by how much was actually achieved. Consider Edison's strategy in terms of Drucker School of Management professor and psychologist Mihaly Csikszentmihalyi's notion of "flow:" In a flow state, time seems to disappear. Concerns about the past or future and other distractions vanish and are replaced by an intense but effortless involvement in the present moment. This present-centered awareness leads to exceptional performance and productivity and is usually experienced as highly pleasurable.

Edison intentionally orchestrated an environment that promoted optimal flow. He eliminated distractions and fully engaged his team in the present moment. A key element of this approach involved balancing intense work with relaxation. As he describes it:

> When experimenting at Menlo Park we had all the way from forty to fifty men. They worked all the time. Each man was allowed from four to six hours' sleep. We had a man who kept tally, and when the time came for one to sleep, he was notified. At midnight we had lunch brought in and served at a long table at which the experimenters sat down . . . and we had a man play the organ while we ate our lunch.

Edison established a unique culture of high intensity and high relaxation. Paul Israel writes:

> Practical jokes, tests of strength and a competition over who could produce the highest voltage with a hand-cranked generator, late-night meals and beer . . . and telling jokes and singing silly or bawdy songs all provided relief

from the pressures of work . . . Edison also [let his staff] use the experimental electric railway that he built in 1880 as transportation to a nearby fishing hole. And workers who lived nearby were free to come and go at the laboratory as long as the work was done.

Through flex time, occasional group outings, midnight luncheons, singing, jokes, and other fun activities, Edison orchestrated a balance between intensity and relaxation that generated an amazing amount of productive energy. Francis Jehl paints a vivid picture of the ambiance of the lab:

> Our [midnight] lunch always ended with a cigar . . . It often happened that while we were enjoying the cigars after our midnight repast, one of the boys would start up a tune on the organ and we would all sing together, or one of the others would give a solo. Another of the boys had a voice that sounded like something between the ring of an old tomato can and a pewter jug. He had one song that he would sing while we roared with laughter . . .

During these midnight lunch breaks, Edison enjoyed "coffee, pie, a cigar, loud music on the organ, and a round of jokes." Adding to the festive nature of these relaxation breaks, former employees as well as personal friends of Edison often dropped by at midnight to partake in these luncheon revels.

As unusual as these activities may seem to modern ears, the acceptance of relaxation as a core part of workplace culture is more important now than ever.

Napping was another of Edison's methods for balancing intensity and relaxation. Despite the rigors of his intense schedule, he required only four to six hours of sleep for much of his life. And, on most days, he indulged in one or two brief naps. By all reports, Edison awoke from these naps without grogginess, ready to plunge into the next task in "full possession of his faculties." Dyer and Martin noted his extraordinary sleep habits while at the West Orange laboratory as follows:

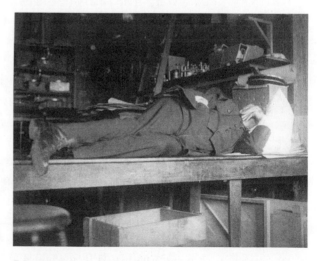

Edison enjoyed taking afternoon catnaps. He used his roll-top desk as a bed until his wife Mina placed a cot in his office.

As one is about to pass out of the library attention is arrested by an incongruity in the form of a cot, which stands in an alcove near the door. Here Edison, throwing himself down, sometimes seeks a short rest during specially long working tours. Sleep is practically instantaneous and profound, and he awakes in immediate and full possession of his faculties, arising from the cot and going directly "back to the job" without a moment's hesitation . . .

Nor were creature comforts of particular importance to Edison when his desire for a nap arose. At Menlo Park, Edison often slept with his head on his chemistry texts, invariably arising with a new angle on how a chemical substance could be used in experimentation. One of Edison's employees remarked to biographers Dyer and Martin about this unusual use the famed inventor made of his books, particularly *Watts' Dictionary of Chemistry:*

Sometimes when Mr. Edison had been working long hours, he would want to have a short sleep. It was one of the funniest things I ever witnessed to see him crawl into an ordinary roll-top desk and curl up and take a nap. If there was a sight that was still more funny, it was to see him turn over on his other side, all the time remaining in the desk. He would use several volumes of *Watts' Dictionary of Chemistry* for a pillow, and we fellows used to say that he absorbed the contents during his sleep, judging from the flow of new ideas he had on waking.

When napping wasn't practical—or when he wasn't tired but just needed to change pace mentally—Edison would simply switch subjects. He said, "In trying to perfect a thing, I sometimes run straight up against a granite wall a hundred feet high. If, after trying and trying, I can't get over it, I turn to something else." When he began to mentally tire, Edison was known to drop what he was doing and begin reading something on a completely different subject. He kept volumes of books on hand for this purpose. Using one activity "as a relief from another" allowed Edison's to keep his mind and body in flow.

Edison's pursuits outside the laboratory also provided an important counterweight to his daily routine. Gardening became a particularly therapeutic form of "loafing," as he affectionately called it. Fishing provided another means for Edison to relax, revitalize, and incubate his ideas. Edison was known to sit for hours fishing with "a baitless hook." Mesmerized by the water, Edison could engage in quiet reflection, allowing his mind to digest the work of the day.

Creating Innovation Literacy: Intensity and Relaxation

> Leisure is the time for doing something useful. This leisure the diligent person will obtain, the lazy one, never.
> —Benjamin Franklin

Shown here in 1909 relaxing on a dock at his Fort Myers, Florida, vacation home, Edison often fished with a baitless hook.

Researchers studying the psychology of memory have discovered that taking regular breaks will improve your recall. If you study something for an hour and then take a ten-minute break, your recall for the material will be higher after the break. Psychologists call this the *reminiscence effect*. Breaks allow your mind the opportunity to incubate and integrate new learning so that your memory functions most effectively. The balance between focused learning and breaks is an example of what high performance coach Jim Loehr calls "oscillation."

Loehr has coached champion athletes including Pete Sampras, Grant Hill, and Michelle Wie, and has applied what he learned from working with athletes to facilitating high performance in the workplace. Loehr emphasizes, "Our most fundamental need as human beings is to spend and recover energy . . . Balancing stress and recovery is critical to high performance, both individually and organizationally."

Edison's "oscillation" between intense hard work and joyful relaxation was a critical element of his remarkable high performance. Many of us spend the majority of our days in a high-stress, driven mode—we're on the phone, at the computer, attending meetings, or rushing to catch a flight. Most people find the ideal of balance attractive, but the reality eludes them because they are just too busy. As the pace of change accelerates, so does the pace of our lives. Death may be nature's ultimate way to of telling us to

Walking was Edison's favorite exercise. Here he strolls with his wife Mina on the lush grounds of their vacation home.

slow down, and nervous breakdowns are frequently the penultimate message. Of course, many more people suffer from mini-breakdowns in the course of the average day: snapping at a coworker, flipping off a fellow driver, or just "zoning out."

You can help avoid breakdowns, large and small, by taking breaks throughout your day. Of course, playing loud music on the organ and smoking cigars at midnight probably won't fit the ethos of most contemporary workplaces, but the principle is the same. Discover activities that help you shift modes and relax. If you discipline yourself to devote ten minutes a few times a day to, for example, practice juggling, yoga or meditation, listen to classical music or jazz, or perhaps experiment with some drawing or journaling, you'll find that you have far more energy and much greater stamina for the intensity of your work.

Moreover, you'll discover that your innovation team meetings become more productive when you incorporate regular breaks. Frequently, an overly intense focus on getting results leads to meetings that go on too long, with an inverse correlation of duration to effectiveness. At the end of a long day without breaks, most people are drained and frustrated. In addition to improving alertness, flow, and recall, regular breaks allow you and your team to take advantage of the mind's natural rhythm of incubation.

POWER NAPPING

Thomas Edison, Albert Einstein, Leonardo da Vinci, Johannes Brahms, Napoléon Bonaparte, Winston Churchill, George Washington, John F. Kennedy, Ronald Reagan, and many other extraordinary achievers, all relied on regular napping to facilitate full engagement. According to the National Sleep Foundation, recent research suggests that napping can boost job performance in a wide range of industries; it improves alertness and memory and can help prevent the fatigue that contributes to many costly mistakes.

Plus, almost everyone has experienced solving a problem by "sleeping on it." This is another example of incubation at work. Creative problem solving, planning, and strategy sessions will be more far more productive if you incorporate an overnight stay. Instead of attempting to get everything done in one intense twelve-hour day, for example, plan a four-hour session on day one, and then a four-hour stint the next morning. You'll find you can get much more done by operating "with the least amount of friction."

ELEMENT 12: SERIOUSNESS AND PLAYFULNESS

> We are most nearly ourselves when we achieve the seriousness of the child at play.
> —*Heraclitus*

Edison's achievement of a balance between the opposite poles of "seriousness" and "play" was a key aspect of his innovation success. As George Parsons Lathrop remarked, "Edison is always himself . . . In the casual, friendly association with him, which good fortune has awarded me, my chief pleasure [has been] the company of one so surcharged as he with what the Norwegians call Liveglade, or 'the joy of life.'"

Although profoundly serious about his work, Edison's playful spirit was always poised to emerge. As Dyer and Martin explain, "Never is he so preoccupied or fretted with cares as not to drop all thought of his work for a few moments to listen to a new story, with a ready smile all the while, and a hearty, boyish laugh at the end."

Edison's laughter, playfulness, and joie de vivre were seamlessly integrated with his dedication, rigor, and seriousness. An afternoon spent solving a knotty, technical problem was, for him, as much fun as an afternoon fishing. Lathrop describes Edison's facility in moving gracefully across the full spectrum of play and seriousness:

He is capable of great jollity and a most charming companionableness. Yet, although he may at one instant be wholly absorbed in a merry chat with friends, laughing at their drolleries and cracking jokes of his own, in the very next instant he will be as completely buried in some abstruse scientific problem as if the conversation had never taken place and the friends had never existed. Neither the friends nor the talk have been ignored, however. It is not indifference which has rendered possible the swift transfer of attention; for his memory, even of trivial details, is extraordinary in its precision and tenacity.

United States presidents, world leaders, and international dignitaries all clamored to meet Edison. As Dyer and Martin describe it, he was constantly barraged

Thomas and his wife Mina were avid gardeners. Edison planted hundreds of varieties of plants, flowers, and trees on his multi-acre Fort Myers, Florida, estate for purposes of research, experimentation, and relaxation.

by heads of state who wanted to be photographed with him, and see him at work in his famous environs: "Celebrities of all kinds and distinguished foreigners—princes, noblemen, ambassadors, artists, litterateurs, scientists, financiers . . . A very large part of the visiting is done by scientific bodies and societies . . . anxious to see everything and be photographed . . . around the central hero."

Indeed, most historical photos capture Edison with hair combed and in formal attire. But, although he could don the attire and play the roles necessary to host such dignitaries, Edison rarely cloaked his sense of humor and playfulness.

One of his favorite tricks while meeting with high-ranking foreign dignitaries, trade delegations, or colleagues was to place white index cards in his suit pockets, each bearing a joke. He'd stealthily draw out a card, tell the joke, then fold and put it away so he wouldn't accidentally tell it again.

Edison confessed his enjoyment of storytelling to a reporter, boasting, "I was very fond of stories and had a choice lot . . . with which I could usually throw a man into convulsions." He loved to share tales, delivering each line with a poker face. Edison's stories had even his devout Methodist mother-in-law, Mary Valinda Miller—Mina's mother—cheering. Straight-laced Valinda found herself "wishing we could tell everything as he can . . ." Edison retained his boyish love of funny stories and pranks throughout his life.

In one hilarious incident, Edison's seriousness collided with practical joking: Intent to discourage those on his laboratory staff who were snitching fine cigars from his private stash, Edison fashioned more than a dozen cigars from barber clippings and shaved cardboard. He hid his good cigars in the photometer room, and placed the phony cigars in a drawer in Francis Jehl's desk so the thieves would have easy access. But, one night Edison was so deeply engaged in a technical problem that he forgot what he had done with the good cigars. Lawrence Frost recounts the result this way:

One evening Edison, while spitting vigorously to get something out of his mouth, remarked, "Francis, those are darn bum cigars you have in that drawer."

Jehl laughingly told him the good cigars were in the photometer room just where he had ordered them placed. Edison replied, "I forgot."

Edison was serious about innovation, so serious that he avoided allowing unnecessary formality to interfere with work. In Edison's lab, every day was "casual Friday." He viewed formal clothing as a confinement that was to be tolerated but not preferred. In adulthood as in childhood, he "never blackened his boots and seldom combed his hair," and even in the laboratory never bothered to don a rubber apron or laboratory coat. Edison's clothing soon displayed evidence of visits to the chemistry laboratory, his pockets coated with chemical powders, and his hands "discolored from some chemical." Although Edison's

Taken during the summer of 1885 or 1886, Thomas and laboratory colleagues Walter Mallory, Charles Batchelor, and Samuel Insull vacation with the Edisons at the Miller family cottage in Chautauqua, New York. Edison reclines in a rocking chair just above the stairs at right, with Mina sitting to his left in a white dress and bonnet. The names of all individuals pictured are listed in the Reference Notes.

first wife, Mary Stillwell, became exasperated with Thomas's tendency to put "all his dirt and grease on my nice white counterpanes [bedcovers]," his second wife, Mina, was more forbearing of his boyish tendencies. When discussing her husband's playful manner at home and at work, she would smile and say, "My job has been always to take care of Mr. Edison—to take care that his home contributed as much as possible to his doing the work that he had to do to the best advantage."

Edison's ability to transcend external formalities, while moving freely through the full spectrum of seriousness and play, endeared him to others. It also offered him access to the childlike openness and creativity that are the hallmarks of genius.

Maturity is often more absurd than youth . . .
    —Thomas Edison

Creating Innovation Literacy: Seriousness and Playfulness

Edison was able to move gracefully and completely along the full spectrum of seriousness and play. He knew that humor was essential to creativity, and that it helped make life worth living. His playfulness allowed him to maintain a broad perspective in the face of intense pressures and to transform stress into energy for high performance, both for himself and his team. The most innovative contemporary workplaces welcome humor and play and the most bureaucratic ones invariably take themselves too seriously. **Overseriousness is a warning sign of mediocrity and bureaucratic thinking.**

As Professor James Clawson of the University of Virginia's Darden School of Business comments,

> An increasing number of case studies support the idea that adding an intentional dose of playfulness to the common mix of seriousness in the business world can produce extraordinary results. From smaller organizations like the Pike's Place Fish Market in Seattle and Menlo Innovations in Ann Arbor, to big companies like Southwest Airlines and Google, a culture that promotes fun and play appears to bring out the best in people at all levels. As Edison understood, people who are having fun are more efficient, productive, and innovative.

No matter where you work, you can benefit from laughing more frequently and fully. Dyer and Martin's description of Edison's laugh conveys a vivid picture of his full engagement: "His laugh, in fact, is sometimes almost aboriginal; slapping his hands delightedly on his knees, he rocks back and forth and fairly shouts his pleasure."

Dr. Madan Kataria is a physician and student of yoga who has studied the benefits of laughter scientifically. He notes the average daily laughter time for healthy humans has declined over the last few generations, from more than twenty minutes a day to less than five. He comments, "This is one of the worst aspects of 'modern life.'"

His findings show that laughter has many profound benefits, including:

- Countering the negative effects of stress

- Boosting the strength of the immune system

- Helping control high blood pressure and heart disease

- Increasing stamina through increased oxygen supply

- Alleviating pain and promoting the feeling of well-being

- Reversing depression, anxiety, and other psychosomatic disorders

Kataria has developed a foolproof method for inspiring deep laughter and has founded more than five thousand laughter clubs and studios in more than fifty countries.

ELEMENT 13: SHARING AND PROTECTING

Edison generated phenomenal energy by finding a balance between intense, hard work and deep relaxation. He managed and transformed stress, for himself and those around him, by balancing his exacting seriousness with "knee-slapping" humor and playfulness. These aspects of his full-spectrum engagement emerged as natural expressions of his character. Edison's character also led him to want to share his ideas openly and freely, but he discovered that innovation success requires a dynamic balance between sharing and protecting one's ideas.

Edison believed that his life's purpose lay in developing innovations that generated the greatest good for the broadest possible audience. As he phrased it: "My philosophy of life is work—bringing out the secrets of nature and applying them for the happiness of man. I know of no better service to render during the short time we are in this world."

From the early days of his career Edison had to deal with threats to his work from would-be competitors who aimed to exploit his achievements. He learned the importance of protecting his ideas and inventions from noted patent attorney, Lemuel Serrell. Serrell urged Edison to file broad patents that "covered the field." This approach provided Edison with significant protection against infringement while also laying the groundwork for continued innovation in each field he desired to penetrate. Some of Edison's patents were filed "to secure if possible the science of the thing"—as was true for many of his electric light, telephone, and phonograph patents—while others were intended to secure commercial processes and applications.

Ultimately, however, Edison believed that "continued innovation" was "the best means of defeating competition." Edison viewed ongoing innovation as his long-term competitive advantage. As Paul Israel comments, "At the core of his strategy was an abiding faith that he could produce technology superior to any competitor's and thus beat anyone in the long run."

Although Edison recognized that protecting his intellectual property was an important priority, he was hesitant to sue those who attempted to infringe. He preferred to devote his time and resources to generating more ideas, improvements, inventions, and innovations. He consistently opposed bringing lawsuits against infringers, saying they

> would require me to give my personal attention to the matter & take me off other far more important work, besides involving us in a great deal of expense & giving our opponents a notoriety which it is hardly desirable they should gain at our expense . . . as long as their work is conducive to their own ruin I see no reason for attacking them.

It was not until 1884—two years after his successful development of a central power station on Pearl Street in New York City—that the number of infringers forced Edison to concede his position. He reluctantly began filing patent infringement suits. Although Edison benefited from patent protection, he became discouraged with the process in the late 1880s and early 1890s after becoming enmeshed in numerous lawsuits. Edison felt that "patent law favored infringers by giving them the benefit of the inevitable delays involved in litigation." He believed the slow rate of prosecution by the courts allowed spurious lawsuits to be filed, with competitors knowing they could market their products until the courts ruled on a victor. As he put it,

> After a thing is perfected and commercially introduced so as to show there is money in it half a dozen parties start to infringe it. The theory upon which they act is, that it will be several years before I can get a final decision, and in the meantime they make money, and when I do get a decision it is probable that no damages can be collected as they have covered their tracks by organizing an irresponsible corporation . . .

He added,

> There is bound to be a delay; why not give the benefit of this delay to the man who has the patent and has worked the thing up to a practical success and not to the infringer who comes after and has no patent. In other words give the man with the patent the preliminary justice until the Court has time to decide the case. Madam Justice has got this thing reversed it seems to me.

In 1910, Dyer and Martin estimated the investment value of industries founded in the United States based on Edison's U.S. patents—excluding the storage battery—to be $6.7 billion, or roughly $100.5 billion today.

Edison had "become extremely skeptical as to the value of any patent, and so long as our patent law remains in its present iniquitous shape, I shall try to do without patents." He began pursuing the use of trade secrets instead, and in the 1890s filed very few patents on his ore-mining technologies. Edison also "particularly objected to the fact that U.S. patents became inoperative before the end of the seventeen-year term if a foreign patent expired first," a legal loophole that "cost him his basic patents on the phonograph and almost claimed his basic lamp patent as well." In 1890, Edison worked with his lawyers to suggest amendments to Congress that would make U.S. patent law more favorable to bona fide inventors, but the House Committee on Patents failed to act on them. Despite his complaints during a tumultuous decade of lawsuits costing Edison millions of dollars, he once again began filing patents in the early 1900s, particularly to protect his storage battery and improved phonographs.

Although he learned to make the most of the patent system—as his record of 1,093 attests—Edison developed other strategies to protect his intellectual property and his reputation in the marketplace. He believed that this could be accomplished by:

- Publishing papers on scientific topics revealing findings related to his laboratory work

- Writing articles for industry magazines and journals about his experiments and invention insights

- Maintaining contact with scientific communities in the United States and overseas

By revealing findings from specific areas of his laboratory research, Edison felt he could broaden the viability of new industries and make them more appealing for investment. Although he did not disclose proprietary information related to his patents prior to their filing, he often shared findings of his experiments. For example, in his early days as a solo inventor, Edison published numerous articles in *The Telegrapher*, an industry journal carrying articles on scientific findings as well as telegraphy inventions. In 1879, Edison published technical information in *Scientific American* about his new high-efficiency dynamo design, which rocked industry perceptions about the amount of power that could be derived from a single machine. Edison's article submissions served not only to share his novel thinking about emerging technologies, they also served to enhance his reputation and stature in the marketplace.

Although Edison consistently positioned himself as an inventor rather than a scientist, sharing his ideas directly with the scientific community became an important part of his innovation success strategy. He appeared before the American Association for the Advancement of Science in 1877, revealing his tasimeter measurements of the sun's corona during a solar eclipse that summer. Two years later, he returned to report his novel findings about occluded gases in platinum filaments, and how his improved vacuum pump "increased [the] capacity of platinum to withstand high temperatures." Edison's strategy of sharing his work with the scientific community paid great dividends. His positive reputation within the scientific world gave him access to prestigious venues including the National Academy of Science in Washington, D.C., and the Royal Society in London.

Edison's commitment to sharing information was also expressed by his funding of the inaugural year of the journal *Science* in 1880, a publication that still exists today. Edison believed the new journal could be an American equivalent of the renowned British journal *Nature*, founded in 1869. *Science* met with an enthusiastic reception, and soon had a list of subscribers including "all the leading universities and colleges" as well as several prominent scientists.

Edison's sharing of his intellectual property outside of the patent framework helped him build his brand and ultimately benefited his businesses. Moreover, he understood that filing patents or using trade secrets wasn't always possible or practical, especially in service and business model innovations. Edison's most profound innovation was his development of the R&D laboratory itself—an achievement for which he never filed a patent.

Contemporary innovators who desire to design proprietary intellectual property beyond the scope of patents can look, as Edison did, to create outstanding processes and methods. No competitor was ever able to duplicate Edison's approach to experimentation, nor his methods of discovery. Edison did not need noncompete agreements because his process and culture of innovation were leagues ahead of potential competitors. He also didn't need formal confidentiality agreements, as his people were remarkably loyal and protective. As Francis Jehl describes:

> These men all had complete faith in his ability and stood by him as on a rock, guarding their work with the secretiveness of a burglar-proof safe. Whenever it leaked out in the world that Edison was succeeding in his work on the electric light, spies and others came to the Park; so it was of utmost importance that the experiments and their results should be kept a secret until Edison had secured the protection of the Patent Office.

The only time that secrets were leaked was when Edison himself was so enthusiastic about something that he couldn't resist sharing it with the press.

Abraham Lincoln, one of Edison's heroes, was a champion of the U.S. patent system. The only U.S. president to apply for and receive a patent (#6469)—for his invention of an inflatable raft to ease boats over sandbars—Lincoln believed that innovation was the quality that distinguished man from the rest of the animals. As he put it, "Man is not the only animal who labors; but he is the only one who improves his workmanship."

Edison struggled with balancing his natural tendency to share his ideas with the realities of protecting them against those who would usurp his efforts. The main difference between Edison's day and the present is that now the usurping happens much faster. As Professor Vijay Govindarajan of the Amos Tuck School at Dartmouth comments,

> The world is full of people who want to commoditize your strategy. Therefore the window in which you can extract profits is shrinking and shrinking. This is why intellectual property (IP) is very critical. Intellectual property *is* the race. It helps create entry barriers. So when you are creating innovation, the more you can create proprietary IP, you can extract profits from your innovation for a longer period of time. What this means is you create intellectual property and continually innovate that IP so you can continue to extract profits. You can't create intellectual property and assume that it will be there forever. But that is the foundation for innovation. And you must continuously update it—that is the key.

Perhaps your natural tendency is to share. Maybe you believe that "information wants to be free," and you are an advocate of the "open innovation" movement, sometimes also called "open source." "Open innovation" is a term coined in 2003 by Henry Chesbrough, now executive director of the Center for Open Innovation at the Haas School of Business at Berkeley. Open innovation is predicated on the notion that there are many smart people in the world who do not work for you. To take advantage of this wider range of potential intellectual capital, many companies are, in effect, "outsourcing" portions of their innovation process. This approach invites external contributions with a view to accelerating the overall innovation process. Some companies, such as Procter & Gamble, Clorox, Kimberly-Clark, Lilly, and Kraft prefer a hybrid approach, with many products innovated internally and others developed through open innovation. Alan G. Lafley, CEO of Procter & Gamble, indicated in 2006 that approximately 20 percent of the company's new products had been generated through open innovation, with a goal of 50 percent by 2010.

There are many ways to approach innovation, but to manage the innovation process effectively, innovation literacy remains imperative. Even firms following an open innovation philosophy must ensure that all their people learn to think like innovators.

And, understanding the nature of intellectual property and how to protect yours, if you choose to do so, is an essential element of innovation literacy.

The challenge of protecting intellectual property has become ever more complex since Edison's time. IBM is at the forefront of the effort to create a new balanced global model for the protection and sharing of intellectual property. Over the last fourteen years, IBM has been the number one recipient of U.S. patents. They operate one of the great industrial labs in the world and do business in 170 countries. Mike Wing, IBM's vice president for strategic communications, explains the company's approach to sharing and protecting. Wing points out that because of pirating, especially in emerging markets, "We've been pushing hard for improving patent quality, for trying to go after patent infringers." Wing adds, "But on the other hand, it's also clear that the model of innovation is changing. It is increasingly collaborative and open . . . A lot of what is being created today is being created in ways that ownership—at best—is widely distributed. And in many cases it's unattributable." He concludes, "In order to maximize innovation, we need to evolve our IP regime, our legal and regulatory regimes, to protect and nurture those kinds of innovations, too. We are looking for—and hope for—a balanced, evolving IP regime on a global basis."

While the biggest players in the world work toward the evolution of a new model for sharing and protecting intellectual property, you can begin by understanding some basics about U.S. patents, copyrights, and trademarks.

Creating Innovation Literacy: Sharing and Protecting

According to the U.S. Patent and Trademark Office (USPTO): "A patent is a property right granted by the Government of the United States of America to an inventor 'to exclude others from making, using, offering for sale, or selling the invention throughout the United States or importing the invention into the United States' for a limited time in exchange for public disclosure of the invention when the patent is granted."

The patent office also makes it clear what can be patented. Utility patents can be granted for:

- Processes

- Machines

- Articles of manufacture

- Compositions of matter

- Improvements of any of the above

Design and plant patents are available for "ornamental design of an article of manufacture" and "asexually reproduced plant varieties," respectively.

The patent office provides the following criteria for inventions to be considered. They must be:

- Novel (from the root *novus* meaning "new"—the core of the word "in*nova*tion")

- Nonobvious

- Useful

- Adequately described or enabled (for one of ordinary skill in the art to make and use the invention)

- Claimed by the inventor in clear and definite terms

The patent office also clarifies what cannot be patented:

- Laws of nature

- Physical phenomena

- Abstract ideas

- Inventions that are not useful (such as perpetual motion machines); or offensive to public morality

- Literary, dramatic, musical, and artistic works (these can be protected by copyright)

A copyright, according to the patent office, is a form of protection available to authors of literary, dramatic, musical, artistic, and certain other intellectual works. Copyright owners have the exclusive right to reproduce and perform their works publicly.

A trademark is, according to the USPTO, "a word, phrase, symbol or design . . . that identifies and distinguishes the source of the goods of one party from those of others." A service mark [SM] is the same as a trademark [TM], except that it identifies and distinguishes the source of a service rather than a product.

Dr. John Wai, a Director of Medicinal Chemistry at Merck & Co., Inc., explains how his team uses international "patent intelligence." They regularly search the patent literature,

> . . . especially in the form of PCT patent applications listed online. PCT stands for Patent Cooperation Treaty. PCT makes it possible to seek patent protection for an invention simultaneously in each of a large number of countries by filing a single "international" patent application instead of filing several separate national applications. Key discoveries in drug design and synthesis usually appear in patent literature months ahead of peer reviewed journals. By carefully analyzing the claims in patent applications, we can gauge the progress of competitors' research. Knowing what is being claimed and what is not is very important. If our competitors are going down the same path we are, we have to file the application as soon as possible or abandon the series. Otherwise we won't be able to obtain the patent coverage to protect our intellectual property, and financially support the project.

ELEMENT 14: COMPLEXITY AND SIMPLICITY

Edison's contemporary, Albert Einstein, said that, "things should be made as simple as possible, not simpler." Edison was a master of this optimal simplicity. He could take his work on the most abstruse technical and scientific challenges and translate his progress into laymen's terms for investors, customers, and the media. He was also able to take extremely complex tasks and set his employees to work with vividly clear, simple instructions. Edison's notebooks are packed with scientific observations revealing high-level thought as well as reams of complex detail. On a given day, he would vault from designing the internal workings of a new electrical power system, or new motion picture projection technology, to working out the precise steps of a manufacturing process.

Edison's ability to discover simplicity in a whirl of complexity was aided by his belief that simple things should remain simple. He insisted that all logistical operations in the laboratory move with precision. Stock rooms were always to be properly filled; chemical and equipment supplies were always to be readily at hand or procured immediately upon request; accurate records were to be maintained and reviewed by laboratory staff at regular intervals; important correspondence was always to be returned in a timely manner. Edison's innovation process required everything to be in clockwork order so that ideas could emerge effortlessly, with no energy wasted on unnecessary pursuits.

Edison set up his laboratory operations to, wherever possible, "get it right the first time." His emphasis on efficiency in the simple things helped make it possible for him to manage complexity.

Edison had the ability to manage many complex endeavors simultaneously. His instinct for simplicity was expressed in his talent for issuing brief, pithy instructions. As Dyer and Martin observed, Edison's instructions were invariably "clear cut and direct." He insisted on uncompromising accuracy in the conduct of experiments focusing on the "minutest exactitude" of every critical detail.

Without precise instructions, experiments might have to be repeated unnecessarily, or the wrong raw materials could be purchased, wasting time and causing missed order deadlines. In one instance, after spending seven hours visiting his new Portland Cement Company factory in Stewartsville, New Jersey, just before the factory was due to begin production, Edison wrote thousands of individual instructions to his staff in the eighteen hours following his visit. As Dyer and Martin note:

When the plant was nearly ready to begin operation, Edison arrived to inspect it. For seven hours he went over it from crusher to packing house, making no notes. On arriving home he stayed up all that night until the following afternoon, writing his suggestions. Some 6,000 were listed and numbered so they could be carried out and reported on by number as they were acted upon.

This may seem like an impossible feat, but Edison had trained himself to process and write information at very high speeds when he was studying to become a master telegrapher, and he retained this ability throughout his career.

Another example of Edison's directness and clarity in delivering instructions can be seen in his exchange with Mr. James Ricalton of Maplewood, New Jersey, the schoolteacher whom Edison selected to search the jungles of the world for a second source of bamboo filament, when he became concerned that Japanese supplies for the first filament might be too rapidly depleted. Ricalton recounts his experience of meeting Edison and receiving his instructions, as follows:

When I was presented to Mr. Edison his way of setting forth the mission he had designated for me was characteristic of how a great mind conceives vast undertakings and commands great things in few words . . . With a quizzical gleam in his eye, he said: "I want a man to ransack all the tropical jungles of the East to find a better fiber for my lamp; I expect it to be found in the palm or bamboo family. How would you like that job?"

Suiting my reply to his love of brevity and dispatch, I said, "That would suit me." "Can you go tomorrow?" was his next question . . . It was while thus engaged that Mr. Edison . . . said: "If you will go up to the house . . . and look behind the sofa in the library you will find a joint of bamboo, a specimen of that found in South America; bring it down and make a study of it; if you find something equal to that I will be satisfied."

Edison's exchange with Ricalton demonstrates his ability to describe a complex task in the simplest and most economical terms. For Edison, brevity was, indeed, the soul of wit.

Edison's capacity to bring clarity to complexity requires what we now call *whole-brain thinking.* His talent for issuing detailed, precise instructions was predicated on his remarkable grasp of the big picture. His associates were amazed and inspired at Edison's comprehensive understanding of the experiments that they were conducting. Dyer and Martin refer to Edison's "instant grasp" of the origin, nature, direction, and interrelationship of a wide variety of projects.

Edison's ability to manage complexity by seeing the big picture as well as the details was more than just a cognitive skill. It was predicated on his remarkable gift for remaining calm and centered while embracing a wide array of challenges. This ability is an essential element in the creation of personal innovation literacy and it's something you can develop.

Creating Innovation Literacy: Complexity and Simplicity

> Nothing is less productive than to make more efficient that which should not be done in the first place.
> —Peter Drucker

If you watch a beginning juggler tossing three balls in the air, you'll notice lots of extraneous movement. Novice jugglers are easily distinguished by, among other quirks, their jaw grinding, chest heaving, and shoulder raising. In comparison, when you watch a professional juggler, the first thing you'll notice is: It looks easy! The professional uses **the right amount of energy in the right place at the right time.**

When he was gearing up the system for incandescent lighting, Thomas Edison was juggling more than forty projects simultaneously. He was a master at investing the right amount of energy in the right place at the right time.

In addition to his whole-brain approach that allowed him to balance his awareness of the big picture with his attention to details and his gift for staying calm and centered in the face of multiple challenges, Edison also mastered another core skill for discovering simplicity in complexity—he knew how to eliminate the unnecessary.

Of course, to eliminate the unnecessary you must have, as Edison did, a very strong awareness of your priorities. As Stephen Covey emphasizes, "Anything less than a conscious commitment to the important is an unconscious commitment to the unimportant."

Here are a few simple things you can do to cultivate this essential element of full-spectrum engagement:

- *Make a mind map of your life purpose and goals and review it on a daily basis.* Then, each day, look at your calendar and consider every thing you are doing in light of your priorities. Focus on eliminating anything that doesn't serve the fulfillment of your goals. At the end of each day, review what you actually did, and again, consider what you might have eliminated. Apply what you learned to plan tomorrow's activities. Rinse and repeat.

- *Cultivate mindfulness.* Edison had a remarkable ability to be fully present in every moment. His "startling . . . power of self-concentration" allowed him to bring "the whole momentum of his mind" to everything he did. This quality of *mindfulness*—bringing full, nonjudgmental awareness to the moment—is something you can learn.

  As Jon Kabat-Zinn, Ph.D., founder of the Center for Mindfulness in Medicine, Health Care, and Society expresses it,

  > Learning to stop, see, understand, and choose are hallmarks of mindfulness and have profound implications for the ongoing development of individuals and organizations. This may seem obvious and simplistic with little application in the complex world of business . . . Yet, our capacity to effectively handle stress, to make informed decisions, and to access previously untapped resources and apply them in challenging, fast-paced business situations all rely on our capacity to be present.

- *Learn to juggle.* Edison juggled multiple projects and one of those projects was to make a movie of juggling. Learning to juggle is a delightful way to develop poise—the ability to have the right amount of energy in the right place at the right time. As Professor Clawson comments, "The best managers are indeed jugglers—their projects are their juggling balls." And, according to recent research by Professor Arne May, of the University of Regensburg in Germany, regular juggling practice leads to "an increase in grey matter in certain areas of the newly trained jugglers' brains."

Joel Jaffe is a game designer at WMS Gaming, an innovative creator of some of the most successful slot-machine games used in casinos worldwide. He describes his Edisonian approach to finding simplicity in complexity:

I work on ten or more game projects at the same time. At first I was concerned that I couldn't juggle so many projects all together. But what I've found is there are patterns in what needs to be done in each one. The more experience you have, the better you know where to put the key emphasis. You don't have to worry about every detail. I also found a creative benefit in handling so much work. I get lots more new ideas and am able to solve problems faster by seeing the commonalities in one set of games, and comparing them to another set. I can use solutions we've generated in one kind of game to create another related solution on a completely different kind of game. It's like having a big bag of tricks.

ELEMENT 15: SOLITUDE AND TEAM

The best thinking has been done in solitude. The worst has been done in turmoil.
—*Thomas Edison*

In his "Talks with Edison," George Parsons Lathrop marvels at "the ease and rapidity with which [Edison] adjusts himself first to one subject and then to another . . ." Lathrop concludes that Edison's flexibility was "due to ready and absolute control of his mental forces." How did Edison cultivate this remarkable control of his mental forces? One of his trademark solutions was to find the optimal balance between working with others and investing in solitude.

Edison functioned as the key idea driver and catalyst behind all the inventions that emerged from his labs, but he relied on his team to translate ideas into innovations. Edison carefully selected employees to support, complement, and expand his thinking. He worked closely with an extraordinary inner circle of collaborators but also engaged on a regular basis with everyone who worked in his labs.

In establishing Menlo Park, Edison organized the entire workspace to facilitate his desire for an easy flow between interaction with others and the need for solitude. He established a private office for himself on the second floor and a "public" office on the ground floor. His upstairs desk provided a haven for his *kaleidoscopic thinking*. His main floor desk served as a place to pay bills, sign payroll checks, and handle correspondence; it was also positioned so that he could greet all the employees as they came and left.

At his West Orange laboratory, Edison further refined his strategy for using physical space to help promote an optimal balance between time alone and engagement with others. He constructed a grand library at one end of the three-story main laboratory building, where he placed his beautiful mahogany office desk. His wife, Mina, also placed a cot in an alcove near the library entrance for his daily "catnaps." This made it easy for Edison to access an ideal environment for his private relaxation and contemplation.

Edison also made special use of "Room 12" in the West Orange facility, an inauspicious room on the second floor near the master chemistry lab. As Dyer and Martin describe it, "No. 12, [is] Edison's favorite room, where he will frequently be found. Plain of aspect, being merely a space boarded off with tongued-and-grooved planks . . . without ornament or floor covering, and containing only a few articles of cheap furniture, this room seems to exercise a nameless charm for him."

Edison used Room 12 to immerse himself in specific questions relating to chemical, mechanical, electrical, or other technical questions. After intense contemplation in solitude, he often invited a handful of employees—or entire teams—to join him for impromptu meetings.

Edison positioned his havens of solitude in proximity to the work area so he could achieve deep states of concentration, yet avail himself of necessary laboratory resources, connect easily with collaborators, and create team interaction as needed. His laboratory colleagues followed Edison's example, working in solitude as well as in partnership, just as Edison did.

Solitude fueled Edison's flow state, and he built all his workplace processes around it. Thus, rather than gauging progress by chronological time, Edison's laboratory teams learned to measure their progress by how close to—or how far from—Edison's own creative output they were running. In an article from the *New York Herald* in January 1879, the reporter describes the back-and-forth dance between Edison and his Menlo Park teams as he fed them ideas generated from his "solitude" hours:

Edison himself flits about, first to one bench, then another, examining here, instructing there; at one place drawing out new fancied designs, at another earnestly watching the progress of some experiment. Sometimes he hastily leaves the busy throng of workmen and for an hour or more is seen by no one. Where he is the general body of assistants do not know or ask, but his few principal men are aware that in a quiet corner upstairs in the old workshop, with a single light to dispel the darkness around, sits the inventor, with pencils and paper, drawing, figuring, pondering. In these moments he is rarely disturbed. If any important question of construction arises on which his advice is necessary the workmen wait. Sometimes they wait for hours in idleness, but at the laboratory such idleness is considered far more profitable than any interference with the inventor while he is in the throes of invention.

By late 1879, however, when the laboratory had to gear up for commercial production of the light bulb, Edison no longer had the luxury of holding his men idle while he pondered in solitude. With more than sixty employees engaged for the scale-up, Edison learned to delegate his work plans more effectively, assigning each detail of the lighting system to a particular staff member or team. Edison continued to work in solitude over portions of the day, but changed the rhythm and timing of his retreats, assigning more managerial tasks to key employees.

Creating Innovation Literacy: Solitude and Team

Where are you physically located when you get your very best ideas? We've been asking this question to people all over the world for the last thirty years. The most common responses are:

"In the shower."

"While resting in bed."

"Driving in my car."

"During a long walk."

It's extremely rare for anyone to say that they get their best ideas at work. What's going on in the shower, bed, highway, or hiking trail that isn't happening in the workplace?

The answer is simple: access to solitude and relaxation. We've already explored the importance of balancing intensity and relaxation for high performance. This balance is complemented with the appropriate oscillation between the stimulation of team exploration and the need for solitary incubation. Finding your optimal rhythm between these two modes will help you be a more prolific, creative thinker and it will also help you liberate more energy for achievement.

Finding balance begins with self-knowledge. Are you more introverted or extroverted? These terms were introduced by psychiatrist Carl Jung and refer to two basic orientations in regard to energy flow. The introvert prefers to focus energy internally and is oriented primarily toward his or her own thoughts. The extrovert's energy flows outward, and prefers to focus on other people and things. Introverts usually need to make more of an effort to engage with other people and extroverts are often reluctant to schedule enough time alone. Whether you are more introverted or extroverted, it's important to discipline yourself to find the balance, as Edison did, between social interaction and solitude. And, this balance is more elusive now than it was in Edison's time. As Peter Suedfeld, Ph.D., a psychologist from the University of British Columbia reports, "My research implies that people are chronically stimulated, both socially and physically, and are probably operating at a stimulation level higher than that for which our species evolved."

In a world of endless meetings, cell-phone calls, BlackBerry messages, and beepers, finding time for solitude requires conscious intention and commitment. **Investment in solitude promotes well-being, creativity, and energy**. And, as Edison understood, it helps us get more out of our social interactions. Psychologists investigating the benefits of solitude, such as Ester Buchholz, Ph.D., author of *The Call of Solitude*, conclude that time alone strengthens our ability to connect with others.

Take time to meditate or just go for a walk by yourself on a daily basis. Once every few months, however hectic your life may be, get away by yourself for at least one day.

When it comes to liberating ourselves from the constant onslaught of noise and other distractions, we all need a little help. One simple aid to the experience of solitude is a pair of noise-canceling headphones. Eliminating sound with noise-canceling headphones is an easy way to get the benefits of solitude even when you're in the presence of others. They're perfect for facilitating a more peaceful inner state on airplanes, trains, in airports—or even right at home.

Another useful, though somewhat unconventional, technology for facilitating the benefits of solitude was introduced by Dr. John Lilly in 1954. The isolation tank—also known as a "float tank"—is a chamber filled with water laden with Epsom salts, where one can rest in comfort free from external distractions and the pull of gravity. It has been used by world-class athletes, including Olympic gold medalist Carl Lewis and members of the Dallas Cowboys football team, to promote well-being and high performance. Nobel Prize winner Richard Feynman was also a fan; he found the isolation tank useful in exploring higher states of consciousness and creativity.

Now known as Restricted Environmental Stimulation Technique (REST), the effects of controlled isolation have been studied at Harvard, Stanford, and many other universities, as well as health-care facilities and athletic training centers around the world. "Floating" has been shown to lower blood pressure, enhance the effects of creative visualization, promote a positive, more optimistic attitude, and even help marksmen improve their aim.

> Be able to be alone. Lose not the advantage of solitude, and the society of thyself.
> —Sir Thomas Browne

The first three competencies for *Innovating Like Edison* place the emphasis on your **intrapersonal intelligence**. In other words, the focus is primarily internal: your mindset, ideation strategies, and energy management all set the stage for the successful application of the last two competencies. If everyone on your team has incorporated the elements of *solution-centered mindset, kaleidoscopic thinking,* and *full-spectrum engagement* then you will find that *master-mind collaboration* and *super-value creation*, which place a greater emphasis on your **interpersonal intelligence**, will be easier to implement.

Chapter Six

COMPETENCY #4— MASTER-MIND COLLABORATION

The word "collaboration" comes from the Latin root *collaborare*, meaning "to labor together, especially intellectually." The term "master mind" was introduced by success expert Napoleon Hill to refer to a very high level of collaboration. He defined it as "coordination of knowledge and effort in a spirit of harmony, between two or more people, for the attainment of a definite purpose." Hill emphasized that when people come together with their passions aligned with common goals, they can multiply their individual intelligence in an expanding framework of positive, creative energy. Hill witnessed the living expression of this idea in the laboratories of Thomas Edison.

The Elements of Master-mind Collaboration are:

16. Recruit for Chemistry and Results
17. Design Multidisciplinary Collaboration Teams
18. Inspire an Environment of Open Exchange
19. Reward Collaboration
20. Become a Master Networker

ELEMENT 16: RECRUIT FOR CHEMISTRY AND RESULTS

Getting a job in Thomas Edison's laboratories in the late nineteenth century was an opportunity comparable to working with Bill Gates or Steve Jobs today. It was a chance to work with the best. Although Microsoft and Apple's hiring processes are a bit more formal than Edison's were, their aim is similar: to find individuals who contribute to the organization's collaborative chemistry.

Edison's wanted people who shared his *solution-centered mindset*. Entrepreneurial spirit, determination, and practical problem-solving ability were more important to him than a résumé. His recruitment tactics were unlike anything conducted in his competitors' laboratories. He required every prospective employee to demonstrate facility with technical knowledge, and the ability to think on their feet, either by conducting experiments or assembling machine parts on the spot without instructions. One of Edison's first hires, John Ott, demonstrated exactly the kind of attitude and ability that Edison sought: "At the age of 21, [John] asked for a job. When an unassembled heap of stock printer parts was pointed to and he was asked if he could make them operate, he replied, 'You needn't pay me if I don't.'" Ott assembled the parts perfectly, and "Edison hired him on the spot, making him a kind of assistant foreman."

Edison's hiring tactics could be likened today to being asked to write an impromptu computer program in Java, or putting up a new Web page in HTML as part of a hiring interview. The bottom line was that, to make it through Edison's hiring gauntlet, you had to have guts, smarts, and a *solution-centered mindset*.

By 1875, Edison had already hired the first four men who would comprise his remarkable "inner circle," including ace experimenter and confidante Charles Batchelor, prototyping expert John Kruesi, lab assistant James Adams, and machinist John Ott. These men all possessed qualities that Edison prized: a broad base of knowledge, a passion for learning, impeccable character, and a commitment to excellence.

When word spread that Edison was hiring, many candidates responded. Prospects seeking to work for Edison at the Menlo Park or West Orange labs made their way in all kinds of weather. Most arrived on the laboratory doorstep without a résumé or letters of recommendation. When your name was called, you'd wipe off your boots and walk into the lab, where you'd be greeted by Edison himself. In addition to the tasks you'd be asked to perform on the spot, Edison asked questions to assess how you might fit with the assignments he most needed. He developed individualized work assignments based on what he required at the moment rather than on a formal job description. Edison wanted people who were able to think for themselves while also following instructions impeccably. He wanted individuals with a broad base of knowledge and a demonstrable hunger for new learning. He looked for individual initiative and independence of mind combined with a willingness to sacrifice for the good of others and put the team first.

When Edison interviewed Reginald Fessenden (1860–1935), for example, Fessenden introduced himself as "an electrician by training." At first, Fessenden protested when Edison said that he wanted to make him a chemist. But then he agreed to take the challenge that Edison presented in his hiring assignment. Fessenden turned out to be an excellent chemist, playing a significant role in the development of insulation for electrical wires. Like many of Edison's recruits, he stayed on for many years, leaving Edison's employ in 1889 to follow his own scientific and business pursuits. Fessenden eventually patented a number of his own inventions and developed the technology to transmit voice and music on the radio. On Christmas Eve 1906, he became the world's first radio performer and DJ, reading selections from the Bible, playing the violin and a phonograph recording of an aria from Handel's opera *Xerxes*. Fessenden claimed that Edison "taught me the right way to experiment."

Edison with his senior team at the West Orange laboratory complex in 1888, after seventy-two hours of nonstop work to complete improvements to a new line of Edison Phonographs.

Edison also sought help finding the right people from his trusted network of stakeholders. In a letter written February 18, 1879, by Grosvenor Lowrey, chief general counsel of Western Union, and a founder of the Edison Electric Light Company, Lowrey recommends that Edison consider hiring Francis Jehl as a laboratory assistant:

My Dear Edison:

Can you make use of a sturdy strong boy about sixteen years old . . . This young fellow . . . named Francis Jehl . . . has a rather awkward appearance, and manners, and is rather slow and might seem to some to be stupid, (but) he is quite an intelligent, industrious, faithful, honest, and high-minded young fellow. He has always been interested in electricity, and while an office boy used to make magnets and little electrical machines which he brought to the office . . .

Yours very truly,
G.P.L.

Edison hired Jehl, and put him to work cleaning and charging battery cells. But he soon proved to be capable of working at a very high level. Jehl played a key role in Edison's breakthrough with the incandescent lamp, and remained an Edison employee, at intervals, for more than forty years.

In 1923, Edison initiated a more formal approach to hiring that included the administration of an extremely challenging "mental fitness test" consisting of fifty questions spanning diverse subjects. Edison modified the tests on several occasions as word leaked out about the answers. They each included a challenging battery of queries, such as the following:

Question: What city in the United States leads in making laundry machines?
Question: Why is cast iron called pig iron?
Question: What telescope is the largest in the world?
Question: Who was Solon?

Edison drew the questions from his own encyclopedic mind, and designed the test to ascertain an individual's breadth and depth of knowledge as well as their potential for "trainability" in his labs. Although he hired more college graduates as his business expanded, Edison wasn't impressed by academic credentials. He once exclaimed: "Men who have gone to college I find amazingly ignorant. They don't seem to know anything." Edison's recruiting and hiring process was usually very successful. Occasionally, of course, some one wasn't "up to snuff." Edison did not tolerate anything less than excellence. The new employee who didn't live up to his expectations was discharged immediately.

Edison was a demanding leader. Although he encouraged an open exchange of ideas and was renowned for his ability to engage in free debate with people at all levels of his organizations, he expected every employee to have "done their homework" prior to voicing an opinion. The man who was unprepared or careless in any aspect of his work received no mercy. Edison paid competitive wages, but money wasn't a prime motivator for him or his staff. Why did so many people want to work for him? Because, as Edison once noted, it was "not the money they wanted, but the chance for their ability to succeed."

Creating Innovation Literacy: Recruit for Chemistry and Results

A few years ago, the leaders of a major pension fund invested tremendous time and energy in crafting an innovative business plan for their organization. They wrote powerful statements of their vision, mission, and values and crafted a compelling strategic plan. Their incentive and compensation packages were competitive, and their training and development initiatives went well beyond anything their competitors were doing. But, there was one elusive issue. As the group's senior leader noted,

> Our hiring process was a reflection of the old model of our business. It focused on bringing us people with great résumés and top-level quantitative skills, but we weren't getting the creative and people skills that we needed to make our vision real. We wanted to put together a collaborative team that was much more diverse, combining quantitative excellence with emotional intelligence and an innovative mindset.

At first, the team considered bringing in consultants to devise a new recruiting and hiring process, but they realized that—with the application of some *kaleidoscopic thinking*—they could do it better themselves.

An internal task force used mind mapping and other *kaleidoscopic thinking* techniques to generate creative ideas on a new process. They then experimented with the evolving process and refined it as a team. As one of the members of the task force commented, "The most tangible benefit from our hiring process, in addition to bringing on board very talented people, was the unifying quality of the collaborative approach. It re-enforced the team alignment around our organizational values, vision and mission with each new hire."

As the folks from the pension group realized, standard recruiting and hiring procedures reflect the fact that "if you do what you've always done you'll get what you've always gotten."

Résumés, a few personality tests, and a couple of interviews with senior managers aren't, as Edison knew, sufficient to find people who will share your team's chemistry and help you get the results you want. Like the pension investment group, Richard Sheridan and his team at Menlo Innovations have created a contemporary version of Edison's hiring process. They call it Extreme Interviewing®; a process with three distinguishing practices:

- *Go beyond standard job interviews and recreate the real working environment.* Just as Edison did, put people in situations that reflect what you really need them to do. Discover how they function in conditions that recreate the actual working environment. As Rich Sheridan comments, "We work hard to align the interviewee's perception of us with our inside reality, which is different than most interview processes, where it is often two to three hours of people creatively lying to each other in order to create a false sense of euphoria about the potential of work and worker. We strive to do just the opposite. We're thrilled if people 'self-select out' during the interview process. We're both better off. Strangely, this makes us even more attractive to the right people."

- *Include the existing team in the process.* If you want a collaborative working environment, then you need a collaborative hiring process. Instead of relying only on the "hiring manager" and a Human Resources representative, get all team members involved.

- *Emphasize hard* and *soft skills.* A *solution-centered mindset*, enthusiasm for learning, and a positive, team-oriented attitude are as important as technical skills. You can get both if you decide not to accept anything less. As Edison contemporary W. Somerset Maugham phrased it, "It is a funny thing about life; if you refuse to accept anything but the best you very often get it." Renowned business consultant Jim Collins, author of *Built to Last* and *Good to Great*, sums up Edison's approach to recruiting, hiring, and firing: "Get the right people on the bus, the wrong people off the bus, and the right people in the right seats."

ELEMENT 17: DESIGN MULTIDISCIPLINARY COLLABORATION TEAMS

What happens when you combine the talents of a British textile machinist, a Swiss watchmaker, an American mathematician with a master's degree in physics, an Irish electrician, a German glassblower, an African-American electrical engineer, and a partially deaf telegrapher? For Edison, the result was a world-beating team of collaborators who churned out hundreds of commercially viable patents and products.

Although Edison was an incomparably brilliant independent innovator, he understood and valued the importance of working with others. He knew he needed a trustworthy team of collaborative employees who could illuminate his blind spots and complement his talents.

Over the course of his career, Edison cultivated an inner circle of roughly ten core collaborators, each contributing materially to the technologies generated by his laboratories. Edison brought together individuals from diverse disciplines whom he would indoctrinate in his methods, then release to freely experiment without his immediate supervision. The diversity of disciplines added tremendous breadth and depth of insight to the laboratory, allowing them to navigate effectively across industry boundaries. The teams were bound together by common values of respect, integrity, and a desire to be the best in the world.

Although Edison was the driver of most innovations flowing from the lab—and was recognized as "the sole directing mind"—most of Edison's work was completed in collaboration with others. Edison was the general manager, coach, and star player of a championship team. He was a master catalyst, motivator, and idea generator; he placed the value of "team accomplishment" at the heart of his laboratory.

Edison assembled teams and designed his organizational structure to complement his own inventing and innovation style. His collaborators had different complementary learning styles and came from diverse disciplines including textile mechanics, drafting, engineering, photography, and mathematics. All were committed to Edison's philosophy of team accomplishment. They shared his *solution-centered mindset* and his core values.

Each member of Edison's inner circle specialized in a particular discipline, but they also were extensively cross-trained. Cross-training enabled Edison to create varying team combinations for conducting the experiments he required, as well as to fill management gaps in his corporate empire. Edison could reach into the inner circle of his organization and combine any two or three of these men with less experienced employees from other parts of the organization, and consistently generate outstanding results.

Edison experimented with various management structures, always focusing on what worked. In his day, there were no formal Purchasing, Operations, Marketing, Information Technology, or Human Resources departments. He created a business model centered around a laboratory environment, offering turnkey technologies and processes ranging all the way from idea development to commercialization, including manufacturing, scale-up, and market launch. In the process, Edison developed the bones of what we see in a modern business operation today. Although few of Edison's staff bore formal titles, we can imagine them fulfilling the following modern-day functions:

Thomas Edison—CEO, President, and Chief Technology Officer

Charles Batchelor—Chief Operating Officer and Vice President of R&D

Edward H. Johnson—Chief Marketing Officer

John Ott—Vice President of Operations

John Kruesi—Vice President of Design and Prototyping

Francis Upton—Vice President of Procurement and Technology

Samuel Insull—Secretary, General Business Manager

Walter S. Mallory—Chief Operating Officer, Ogden Ore Milling and Portland Cement Works

Harry F. Miller—Secretary and Treasurer, Edison Phonograph Works (no relation to Mina Miller)

Frank Dyer—General Counsel

Edison's collaborative approach clearly differentiated him from other inventors of his day, as Dyer and Martin explain:

In this respect of collaboration, Edison has always adopted a policy that must in part be taken to explain his many successes. Some inventors of the greatest ability, dealing with ideas and conception of importance, have found it impossible to organize, even to tolerate a staff of co-workers, preferring solitary and secret toil, incapable of team work, or jealous of any intrusion that could possibly bar them from a full and complete claim to the result when obtained. Edison always stood shoulder to shoulder with his associates, but no one questioned the leadership, nor was it ever in doubt where the inspiration originated.

Edison's formidable inner circle of collaborators was supported by a "second circle" composed of fast-track employees and other stakeholders. The second circle played an important role in the success of the laboratory, ensuring that Edison always had a deep bench upon which he could draw. Members of the second circle represented an important "go to" network when inner circle members were called away to other regions of Edison's empire. The second circle ran much of the day-to-day manufacturing and customer service functions; it offered a mechanism through which Edison could easily reach the edges of his organization. In twenty-first-century terms, Edison's second circle allowed him to be only "two clicks away" from communicating with any employee in his operation.

Key members of the second circle not employed by Edison's laboratory or manufacturing operations functioned much like a trusted board of directors might operate today. This group included business mentors, professional specialists, and providers of various forms of financial, legal, or scientific support to Edison's laboratory.

The instructive element lies in understanding how the inner and second circle all supported Edison's passion for innovation, and the way that Edison orchestrated their diverse talents and backgrounds to create his collaborative "master mind."

Recent research confirms the wisdom of Edison's approach to collaboration. In *Notebooks of the Mind*, University of New Mexico Professor Vera John-Steiner states that, in all true collaborative efforts there is a "weaving together of ideas, styles of work, and approaches that characterize different disciplines . . ." The collaboration itself is a means to compensate "for each other's blind spots." "Collaboration operates through a process in which the successful intellectual achievements of one person arouse the intellectual passions and enthusiasms of others . . ." It involves "visualizing a solution by working with ideas that cut across disciplines . . ."

John-Steiner emphasizes that the process is facilitated by assembling teams with members of differing age and experience levels who bring a diversity of ideas and problem-solving approaches. She also notes the positive role of visual images and three-dimensional models to represent concepts.

Edison intuitively applied all the elements necessary to optimize team performance. His approach demonstrated how "the successful intellectual achievements of one person arouse the intellectual passions and enthusiasms of others." As John-Steiner puts it, by forming diverse, multidisciplinary collaboration teams, and "visualizing solutions that cut across disciplines," Edison was able to generate more innovation than anyone else, and create an organization flexible enough to manage the diverse business models that flowed from their unprecedented success.

Creating Innovation Literacy: Design Multidisciplinary Collaboration Teams

IDEO is the award-winning design firm responsible for helping to develop innovations ranging from the Palm V PDA, to Steelcase's Leap office chair. In addition to helping clients create innovative products and services, IDEO has grown a successful business "enabling organizations to transform their cultures and build the capabilities required to innovate routinely."

What is the key element in IDEO's approach to innovation? As their Web site states: "Multidisciplinary teams are the heart of the IDEO method. It's no accident. We believe this is how innovation happens in the world."

IDEO's four hundred-plus staff includes individuals from a wide range of backgrounds including anthropology, computer science, engineering, graphic design, health care, and psychology. Tom Kelley, IDEO's CEO, has written a book entitled *The Ten Faces of Innovation* in which he describes his approach to crafting multidisciplinary collaboration teams. In addition to actively seeking individuals from diverse technical backgrounds, Kelley emphasizes the ten "faces" or roles that he looks for to create master-mind collaboration within his teams.

You can complement your study of the "ten faces" with another extremely valuable approach to identifying, coordinating, and working effectively with different types of people. The Enneagram is a ninefold typology that illuminates people's different core motivations, approaches to problem solving, and styles of communicating. The system provides valuable insights into how the different types function under stress and how they can evolve to be their best. It is a valuable tool for discovering your own strengths and areas for growth while also helping you learn how to work more effectively with others. Boeing, Kodak, Hewlett-Packard, Toyota, Sony, and many other organizations have utilized the Enneagram to help structure and develop their collaborative team efforts.

If, like economist Kenneth Boulding, you believe there are only two types of people—those who divide everything into two groups, and those who don't—then perhaps a ten- or ninefold system seems too complex. The good news is there's another simpler but very useful tool for crafting balanced collaboration teams. Created by Ned Hermann, a former management educator at General Electric, the Hermann Brain Dominance Instrument (HBDI) discerns four different thinking style profiles. Organizations including IBM, American Express, and Target build "whole brain teams" using this approach. As we've seen, Edison's "whole-

Inductees of the National Inventors Hall of Fame on multidisciplinary collaboration teams:

> As technology gets more complex, it is absolutely necessary to have multidisciplinary teams because, in many cases, one person can't retain all the information that's necessary to make breakthroughs.
> —Dr. Jim West, 1999 inductee, National Inventors Hall of Fame

> My lab has people with 10 to 12 different disciplines in it—molecular biologists, cell biologists, clinicians, pharmacists, chemical engineers, electrical engineers, materials scientists, physicists and others. Many of our ideas—such as tissue engineering—require these different disciplines to move from concept to clinical practice. It makes it possible to do nearly anything "discipline wise" in the lab.
> —Dr. Robert Langer, 2006 inductee, National Inventors Hall of Fame

Langer also comments on the importance of nurturing, as Edison did, a "second circle":

> There are a few senior people I rely on, but we have about 30 people with Ph.D.s or M.D.s and another 30 graduate students who report to me. We have educated the next generation of biological engineers. Over 150 trainees from our laboratory are now professors training others in this area; an equal number have started or are now working in biotech or medical device companies.

brain thinking" was an important element in his success. He crafted teams that reflected this balance and this approach is now, more than ever, a critical element in assembling a *master-mind collaboration* team.

ELEMENT 18: INSPIRE AN ENVIRONMENT OF OPEN EXCHANGE

In his classic book *The Fifth Discipline*, Peter Senge introduced the notion of "the learning organization," which he describes as a place "where people continually expand their capacity to create the results they truly desire, where new and expansive patterns of thinking are nurtured, where collective aspiration is set free, and where people are continually learning how to learn together." Senge's description touches the heart of how Edison shaped collaborative environments within his laboratories and manufacturing operations.

Edison envisioned both Menlo Park and West Orange as "campuses" where a social fabric of collaboration could be created and nurtured. He designed the physical layout of his laboratories to aid the rapid diffusion of "new and expansive patterns of thinking" with work areas placed adjacent to meeting areas. At both facilities, the central buildings were flooded with natural light from large windows spanning all four sides of the outer walls. Flowing natural light helped create a positive and pleasant working atmosphere. The unique layout of the labs became a major drawing card for the top talent Edison sought.

The inspired, open environment of exchange within the lab constituted the heartbeat of Edison's innovation empire. As noted by historian Andre Millard, who studied the West Orange operations in detail, "The laboratory initiated the work of the greater organization rather than responded to its needs." And Edison's larger organization became renowned as a place where people could "continually expand their capacity to create the results they truly desire."

Although the cutting-edge equipment and novel architectural layout of Menlo Park and West Orange contributed significantly to an environment of open exchange—"where people are continually learning how to learn together"—it was Edison himself who provided the human spark. He exuded a compelling intellectual effervescence that motivated people to exchange ideas openly. Edison's love of learning infused his laboratories, creating an atmosphere that promoted idea exchange. Menlo Park and West Orange were designed so that Edison could circulate through the labs with ease, making it much easier for him to give and receive feedback. Over time, Edison worked shoulder-to-shoulder with every member of his laboratory staff. He valued ideas arising from all levels. He also encouraged intellectual exchange with the many influential visitors who came from all over the world to visit his facility.

Edison created a vibrant think tank where employees were free to generate, share, and

This February 1880 shot of the second floor at Menlo Park shows Edison wearing an artisan's cap and kerchief, sitting in front of the pipe organ often used for entertaining his staff during "midnight lunches." Incandescent lights hang from the gaslight fixtures on the ceiling. Names of employees shown are listed in the reference notes.

then test their ideas. Edison was the ultimate decision maker, but he knew his decisions would be better if he was informed by a full spectrum of discussion. He encouraged lively, respectful debate, entertaining a wide range of conflicting viewpoints from all levels of his team. Edison successfully facilitated a true democracy of ideas. He created a level intellectual playing field where there was no pecking order to poison exchanges. As Dyer and Martin describe it:

> Those who were gathered around him in the old Menlo Park laboratory enjoyed his confidence, and he theirs. Nor was this confidence ever abused. He was respected with a respect which only great men can obtain, and he never showed by any word or act that he was their employer in a sense that would hurt their feelings, as is often the case in the ordinary course of business life. He conversed, argued, and disputed with us all as if he were a colleague on the same footing. It was his winning ways and manners that attracted us all so loyally to his side, and made us ever ready with a boundless devotion to execute any request or desire.

Contemporary sociological and psychological studies demonstrate consistently that the collaborative, open model developed by Edison optimizes the confluence of creativity, strategy, and action. As stated by Vera John-Steiner, the "fluency of ideas and flexibility of approach characterizes scientifically creative individuals working together on a problem."

Complexity experts Welter and Egmon emphasize that the most productive collaborative environments are characterized by:

- The freedom to step out of mainstream thought

- Commitment to shared values

- Genuine curiosity about possibilities and opportunities

- A compelling shared desire to improve the current state of quality, efficiency, and/or technology

- Genuine self-awareness of strengths and weaknesses in learning styles.

These characteristics describe Edison's labs. The atmosphere of open exchange served to empower his staff. As Edison's businesses expanded, he taught his teams to operate without him for extended periods of time. Although he provided initial guidance and suggestions on how to approach problems, he encouraged his experimenters to find their own solutions. As he noted,

> I generally instructed them on the general idea of what I wanted carried out, and when I came across an assistant who was in any way ingenious, I sometimes refused to help him out in his experiments, telling him to see if he could not work it out himself, so as to encourage him.

With more than 300,000 employees globally, IBM CEO Sam Palmisano faces a profound challenge: how to inspire an open environment of exchange with every employee, enabling each individual to feel his or her input is valued.

Palmisano found a solution in the company's program of online "jams"—massive, global events that run from forty-eight to ninety-six hours in length, each organized around a specific theme. Designed to allow unfettered dialogue across all parts and levels of the organization, IBM's jams openly display employee comments—as well as Palmisano's—on themes ranging from corporate values to new business ideas. The inaugural event, WorldJam, took place in 2001, sponsored by Palmisano's predecessor, Lou Gerstner. The company has since conducted six more jams within IBM. The jams have met with such internal success that IBM has begun helping its clients develop their own jam programs.

IBM's jam themes have focused on managers, consultants, and even the company's core values. During the two parts of 2006's InnovationJam—the first part to surface opportunities, and the second to hone them as businesses—more than 140,000 IBMers, their family members, clients, and business partners from 104 countries came together online to generate more than 46,000 ideas for new products and services. At the end of the second stage, $100 million was set aside to fund the ten best ideas. As Palmisano notes, "Collaborative innovation models require you to trust the creativity and intelligence of your employees, your clients, and other members of your innovation network. We opened up our labs and said to the world, 'Here are our crown jewels, have at them.' The jam, and programs like it, are greatly accelerating our ability to innovate in meaningful ways for business and society."

Edison balanced his desire for his people to imitate specific aspects of his trade-mark work style with a sincere desire for each individual and team to discover their own methods. Thus, he succeeded in setting free their "collective aspiration" in a way that promoted an astonishing level of practical innovation.

Creating Innovation Literacy: Inspire an Environment of Open Exchange

Edison's charismatic optimism, passionate curiosity, love of learning, storytelling, and appreciation for the diverse talents of his staff all contributed to the creation of an unprecedented culture of innovation. The environment of open exchange that he encouraged was an essential ingredient in sustaining his organization's success. *Appreciative Inquiry* is a contemporary organizational development methodology that can help you facilitate an environment that supports this critical element of *master-mind collaboration*.

Most organizational development initiatives are based on the apparently reasonable assumption that it's important to focus on what's wrong in order to fix it. They often begin with an extensive survey and problem analysis followed by a diagnosis of strategic, operational, and cultural dysfunctions. Appreciative Inquiry is predicated on a radically different assumption. It begins with the notion that the process of facilitating an open dialogue about dreams, hopes, visions, values, successes, and strengths is, in itself, transformational.

As Dr. Diana Whitney, founder and president of Corporation for Positive Change explains, Appreciative Inquiry works "by focusing the attention of an organization on its most positive potential—its positive core. The positive core is the essential nature of the organization at its best—people's collective wisdom about the organization's tangible and intangible strengths, capabilities, resources, potentials, and assets."

Appreciative Inquiry unfolds through a process known as the 4D Cycle. The four phases of the cycle are Discovery, Dreams, Design, and Destiny.

Discovery: The Discovery phase begins with the collaborative creation of open-ended questions focused on illuminating organizational core strengths. As stories of best practices are shared, the collective wisdom of the organization comes into clearer focus. In the process, new relationships and cross-functional alliances are formed, and, as Whitney emphasizes, this often inspires the spontaneous emergence of changes that promote a more open environment of exchange.

Dream: The Dream phase involves organization-wide *kaleidoscopic thinking* devoted to nurturing images of a positive future. It usually includes a series of large-group, cross-disciplinary, multilevel forums focusing on sharing innovative strategic visions.

Design: The Design phase gets people at all levels of the organization involved in aligning the positive passions and competencies identified in the Discovery phase, with the visions and goals that arise through the Dream phase. It involves the articulation of statements of the ideal organization, clarifying how people want to work together on a daily basis to make their ideals real.

Destiny: Although positive change tends to emerge spontaneously at every stage of the Appreciate Inquiry process, the Destiny phase focuses on formulation and implementation of specific individual, team, and system-wide commitments and actions supporting further continuous learning and innovation.

Appreciative Inquiry works because it elegantly uses open-ended questions and open forums to help organizations become more communicative. It marries the fundamental human need for appreciation with our natural aspiration for growth and learning. Appreciative Inquiry has been applied with very positive results at corporations including British Airways, GE-Capital, Merck, Verizon, and Sandia National Laboratories, as well as a many nonprofit, social service, and government organizations.

Steve Odland, chairman and CEO of Office Depot, comments on creating an environment of open exchange:

"Edison was a master at figuring out ways to address resistance while creating high-efficiency circuits. In an organization, you've also got to address resistance—by removing it—and the biggest resistance in a company is 'fear.' And you've just got to take the 'fear' out of it, and allow people to be themselves—to come up with ideas and try different things, and not worry about negative repercussions—and only face positive repercussions, which then creates speed."

ELEMENT 19: REWARD COLLABORATION

The best reward is not to give money but to show a new connection toward the future.

—*Dr. Clotaire Rapaille, Seven Secrets of Marketing*

Although Edison made a fortune from his work, money was never his main motivation. He loved creating "a new connection toward the future," savoring every moment of the process of invention and innovation. And, of course, he recruited people who shared this perspective. By sustaining an environment of open exchange, continuous learning, and commitment to excellence, Edison created a working environment that was intrinsically rewarding for his staff. Moreover, as his fame grew, talented individuals recognized that the chance to work with Edison was a priceless, historic opportunity.

The opportunity to work with and learn from Edison was also empowering. He helped his people discover their strengths and capitalize on them. As John Ott, who was originally hired as an assistant, said of his employer, "Edison made your work interesting. He made me feel that I was making something with him. I wasn't just a workman."

Ott's name eventually appeared with Edison's on various patents for the incandescent lighting system, and he received a 50 percent share of royalties accruing from patents bearing his name.

The learning experience Edison offered was greater than anything his competitors, or the universities of his time, could muster. As Reginald Fessenden, the electrician-turned-chemist, described it, those men who "had been 'indoctrinated' in his methods" were given "a very considerable opportunity of developing their individuality in the working out of [new kinds of] problems." By soaking in Edison's methods "after working under close instructions from Edison" most workers "got to understand his methods pretty well and seldom needed to bother him about details." And, like Fessenden, many achieved great success during and after their tenure at Menlo Park or West Orange by bringing the unique learning gained from Edison to their new work environments.

Edison encouraged all staff members to take advantage of the seminars he personally delivered from the foyer of the West Orange laboratory. These weekly presentations featured Edison's best thinking on a variety of scientific and technical subjects. His staff realized these seminars were a golden opportunity for learning and advancement, and they were always well attended. The staff viewed the opportunity to learn directly from Edison as a special reward for their work. And, as a result of their learning, they worked more effectively, thereby gaining more access to the financial rewards Edison offered.

Edison consistently encouraged and rewarded self-education for all laboratory employees, particularly at West Orange, giving them access to his outstanding private library, "one of the finest scientific and technical libraries in the world." Reginald Fessenden recalled that he and Arthur Kennelly "used to study mathematics together during lunch and that we often spent an hour or so studying theoretical physics or chemistry at the end of the day."

Edison also offered financial incentives for those who showed the kind of entrepreneurial spirit he prized. His new employees usually received detailed written instructions that could keep them very busy for a long time. But he always encouraged those who took extra initiative to go beyond what was prescribed. As Paul Israel comments, "The laboratory's machinists could . . . add to their earnings through inside contracting in which they bid on work within the shop for the construction of experimental apparatus, demonstration or manufacturing models, and special tools for manufacturing them. Taking on such a contract required initiative that might lead to other responsibilities and possibly a supervisory role in Edison's factories."

Although Edison was a great believer in rewarding individual initiative, he always sought to leverage individual incentives for maximum benefit to the team effort. He offered the richest financial rewards to his right-hand man Charles Batchelor, who received 10 percent of the profits from Edison's inventions. Beyond his exceptional skill in the laboratory as an experimenter, Batchelor in essence served as Edison's chief operating officer and vice president of R&D. A brilliant manager, Batchelor played a critical role in assembling and bringing out the best in all of Edison's teams. Edison spawned more than 150 businesses, and he offered key team members the opportunity to "get in on the ground floor" by purchasing stock. His most generous offers went to those who helped manage the innovation process as well as those whose individual brilliance helped to drive it. Inner circle members who contributed to the management of Edison's burgeoning empire all became wealthy as a result.

Those who were rewarded for individual brilliance included William Kennedy Laurie Dickson, who contributed foundational insights to the moving picture technology he and Edison developed. Dickson received a percentage of the royalties accruing from the kinetoscope and the motion picture camera. James Adams, an early member of Edison's Ward Street laboratory in Newark, stayed with Edison through the Menlo Park years, and was also given a significant royalty in recognition of his contributions to the development of the electric pen, phonograph, and other lucrative innovations. Francis Upton, the precocious mathematician and physicist who assisted Edison in developing reams of detailed market analyses and risk profiles for the electric light, received 5 percent.

More individuals would have received royalties if their names had been assigned as co-inventor on one of Edison's hundreds of patents, but Edison chose not to pursue this path partly because existing patent laws made joint patents more vulnerable to challenge.

As Edison's role expanded from running a moderately sized laboratory at Menlo Park to running a massive industrial R&D operation at West Orange, including a large manufacturing and corporate empire, he required a larger supporting cast. And, as his second circle expanded, Edison facilitated their development through rewards of "position." By serving as well-paid assistants to Edison's inner circle staff at domestic or international trade shows, they enjoyed opportunities for travel, and enhanced prominence in their field. Second circle members showing special promise were also given opportunities to serve as temporary replacements for inner circle members away on special assignments. These replacement stints prepared second circle members for eventual inner circle membership.

Beyond the first and second circle, everyone in Edison's organizations believed that his good work would be recognized and rewarded. As Edison once explained to a reporter, his people "know that if I am successful that I don't keep it all for myself." Edison paid his machinists competitive wages and he sought to appropriately compensate all those who made crucial contributions to his innovations.

Edison carefully leveraged individual rewards for his people at all levels to support team accomplishment. He created a unique, empowering, learning environment that inspired his diverse staff to work together. The collaboration he generated yielded unprecedented results and serves as an inspiring model for innovation today.

Creating Innovation Literacy: Reward Collaboration

Edison understood that the experience of positive collaboration is intrinsically rewarding. He knew by creating multidisciplinary teams and inspiring an environment of open exchange, his people would experience a "new connection to the future." Over the years at both Menlo Park and West Orange, many individuals volunteered to work for Edison for months without pay, just to have the opportunity to learn from him and be part of the collaborative experiences.

At SRI International, Carlson and Wilmot emphasize: "Rewards come in many forms. A primary reward can be the opportunity to work on a terrific project with wonderful colleagues."

The most innovative organizations provide individuals with a feeling of being challenged and supported to do their best. Edison gave his "Muckers" the feeling they were important contributors to his team. He demonstrated his respect by "turning them loose" and expecting great things from them, and he usually wasn't disappointed. Financial benefits were, of course, part of the overall reward for Edison and his teams. He faced the challenge of rewarding exceptional individual achievement in a way that didn't distract from the emphasis on collaboration. Of course, over the years, some of his people felt they deserved more money and recognition than they received, but overall Edison did an amazing job of managing the process. His staff was remarkably loyal, and the vast majority believed their good work was recognized and rewarded appropriately.

In our combined sixty years of working with organizations, we've witnessed many clients invest tremendous time and energy in crafting complex incentive and compensation plans. Sometimes they succeed in designing rewards that reinforce the behaviors they seek; and in many cases, they spend lots of money to reinforce the opposite of what they hope to achieve. We don't specialize in incentive and compensation consulting but we have learned one very important point about rewarding collaboration to promote innovation: **To reward collaboration be sure that you are not punishing it.**

A former client and good friend of ours was recently hired as president of a rapidly growing engineering firm. The chairman had hired him because of his proven success as an innovation leader in his previous positions in the industry. He describes the challenge he faced:

> As the company had been experiencing rapid growth over the past five years, certain plans put into place now required some serious overhaul. Growth also meant there was now a much stronger need for collaboration between offices and amongst service lines.
>
> While getting to know the office leaders and learning how they operated, I soon came to realize how almost every decision they made was tied into their incentive compensation plan. One of the "leaders" actually told me that if I asked him to something that affected his bonus, he would refuse to do it.

The incentive plan, I discovered, was an exact, complex formula that only an engineer could figure out. The calculations were primarily based on individual performance. It did help bring about desired results when the company was smaller and less dynamic. But, the problem now is that decisions affecting increased growth of the business, such as investments in training, new markets and expanded services, etc. are colored by their affect on short-term profitability of the individual operations. Thus, in certain ways, the parts have become greater than the whole.

My challenge is to keep the good components of the old plan while phasing out elements that interfere with the collaborative, innovative approach we need to take full advantage of our growth opportunities.

In addition to eliminating disincentives, it's important to discover your own creative ways to reward collaboration and promote innovation. Dr. Annaliesa Anderson, a Senior Research Fellow in Vaccine and Biologics Research with Merck & Co., Inc., describes her creative initiative to reward collaboration:

When I joined the pharmaceutical industry, I remember noticing that what often took the most prominent position on peoples' desks was an acrylic "cube" containing a small drug vial. These cubes represented the contribution that individual had made in developing a new therapeutic drug. In our industry, we have many more misses than hits, so these vials are few and far between.

My team invented a vaccine that passed all its preclinical tests and was destined for clinical trials. The teamwork required to achieve this milestone was monumental. One person came up with the core idea, but it was our team who refined it and made it work. So the question became, "How can I reward my teams' efforts?" I opted for "the cube." It was something I had always prized, and it serves as an ongoing symbol of the greatest success we can hope for in our industry: developing drugs that improve peoples' lives.

My original plan was to use a cube to reward my immediate team of about 10 people. But I also decided to reach out to other multi-disciplinary teams who'd assisted us in reaching clinical trial, including Biology, Engineering, Clinical Sciences and Marketing.

I was stunned that everyone got on board and wanted to include their groups. We ended up distributing 150 of the cubes at a private meeting, with cake for everyone. The team leaders for each functional area showed slides naming all the people on their team, and what their contributions were. It was great! It didn't cost much, but had huge impact. Now, when I visit peoples' offices, our cube is there, prominently displayed on their desks. Mine is on a shelf—otherwise it would get buried in all the papers on my desk!

Mike Wing, IBM's Vice President of Strategic Communications, sums up the challenges and the promise of nurturing a culture that rewards collaboration:

We're on a journey toward fully rewarding collaboration . . . we have a lot more evolution to do in terms of re-thinking our reward structure, our career paths, our collaboration models, to be optimized for innovation . . . But one thing you *can* do is reward people with the freedom to express themselves—engage, collaborate. In other words, in a certain sense, virtue is its own reward. Innovation is its own reward.

If you come to a company like IBM, you probably come because you hope to make a difference, because the place tackles big, difficult, meaningful problems and issues. We don't make gizmos. We map the genome, we send people to the moon, we move individual atoms, we tackle modeling epidemiology.

So in a way, you could say that the reward system is: *not whacking people* . . . more and more, it's also about building a culture that gives permission, and isn't dominated by fear but rather by hope.

ELEMENT 20: BECOME A MASTER NETWORKER

Edison's approach to facilitating *master-mind collaboration* began with his focus on recruiting highly motivated people who could get results. Then, he organized them in multidisciplinary teams and encouraged the free flow of ideas while providing a range of rewards and incentives for collaborative behavior. Though all these elements optimized his internal intellectual capital, Edison also knew that the most productive thinking required a steady stream of contribution from a wide range of resources outside his laboratory.

Beyond the world of beakers and test tubes, Edison realized the importance of maintaining a vibrant connection with the diverse constituencies fueling his innovation empire. He cultivated relationships with people from a broad spectrum of disciplines, including technical experts, customers and prospects, journalists, academics, financiers, and politicians.

Edison's gifts of wit and showmanship distinguished him from his competitors, most of whom were introverts focused exclusively on technical issues. His networking efforts flowed easily and naturally, yet were laser-targeted. Although he frequently corresponded with contacts through handwritten letters or telegrams, Edison knew the importance of "taking the pulse" of his network through personal face-to-face meetings. He cultivated relationships that served his evolving master mind, and facilitated access to investment and promotion. He also leveraged large-scale networking venues—such as major domestic or international trade shows—to full advantage.

Edison at age seventy-four (second to left) reads a newspaper while networking with Henry Ford (far left), President Warren Harding (second from right), and Harvey Firestone (far right) during a 1921 summer camping trip in the mountains of Maryland.

Edison understood that networking is different from marketing or public relations. While all three of these tools target key influencers, Edison knew his networking efforts generated the most personal, in-depth information exchange, particularly at high levels. Moreover, he recognized that marketing, which generally focuses on reaching large, targeted groups of consumers or business people with a pre-established message, and public relations, which typically involves placing messages in a large-scale environment with the intention of reaching a specific target audience in ways that encourage rapid diffusion, were both done more effectively when they were informed by the results of networking. Edison was a pioneer in marketing and public relations, and his successes in these areas were fueled by his networking prowess.

Although the term "networking" did not exist in Edison's day, he knew instinctively how to do it, and how crucial it would be to his success over the years. His networking strategy involved the following actions:

- Carefully research industry trends, then seek out and cultivate relationships with leading experts in key areas of interest.

- Cultivate deep, long-term relationships with the most knowledgeable and influential people in a wide variety of fields, especially publishers and reporters.

- Seek out personal interaction, either one-on-one or in small groups.

- Participate in large, high-visibility trade shows in the United States and abroad.

- Help others achieve their goals. Edison thought carefully about the needs of those he met, and how he could help them.

- Prepare a brief statement describing what is desired from each contact.

- Manifest charismatic optimism complemented with wit, humor, and a flair for storytelling, inspiring people to remember you.

Edison employed this networking approach throughout his career, from his days as a small businessman (1863–78) through his tenure as an internationally famous mogul (1878–1931). Edison moved from regional to national to international fame with relative ease, and refined his approach to accommodate a growing network. As his fame grew, so did the flair, drama, and panache with which he captivated his diverse audiences.

Edison's natural gift for networking emerged in his preteen years when he cultivated the relationships that helped him develop skills to move rapidly through the ranks of telegraph operators. Edison bonded with other roving teenage telegraphers, maintaining relationships with many that lasted for decades. Ezra Gilliland, for example, was a fellow "lightning jockey" who became a wealthy business owner in Cincinnati. Edison nurtured the connection with his old friend, and Gilliland ultimately played a significant role in helping Edison launch the electric pen as well as other inventions.

Edison's daily newspaper reading kept him abreast of market trends, enabling him to develop target lists of new contacts, especially those who might serve as sources of funding for his inventions. A pioneer of the form of networking we now call "lobbying," Edison noted he "kept posted, and knew from their activity every member of Congress, and what committees they were on; and all about the topical doings."

Edison's experience working as a newsboy, reporter, and press copywriter also fueled his passion to network with individual publishers and reporters. He developed relationships with several journalists who became important to his career. One media connection Edison carefully nurtured proved critical to his success. From his earliest years as an inventor, Edison had cultivated relationships with key individuals at *Scientific American* magazine. Edison was thus able to obtain a high-level meeting on short notice with its editors in December 1877, just days after completing the first working prototype of the phonograph. Edison and Charles Batchelor presented the new cylinder phonograph in the publisher's New York offices:

> There they amazed the staff by playing the little machine on the editor's desk and turning the handle to reproduce a recording he had already made. According to the journal's editor, "the machine inquired as to our health, and asked how we liked the phonograph, informed us that it was very well, and bid us a cordial good night. These remarks were not only audible to ourselves, but to a dozen or more persons gathered around . . ."

The enthusiastic reception by the editors at *Scientific American* initiated a cascade of interest in Edison's invention and the story of the phonograph began buzzing through the country, beginning regionally in New York and New Jersey then spreading down the east coast to Washington, D.C., where Edison received requests to demonstrate the phonograph not only for Congress, but for President Rutherford B. Hayes himself. Soon, influential reporters from the *Boston Globe*, the *New York Sun*, *Harper's Weekly*, and many others began contacting Edison, requesting interviews. Within months, the *New York Daily Graphic* had dubbed him "The Wizard of Menlo Park"—a nickname that endured throughout his career.

The reporters whom Edison befriended during the early frenzy over the phonograph became important members of his network over subsequent years. These included W. A. Croffut (*New York Daily Graphic*) and Amos Cummings (*New York Sun*), who both offered valuable advice to Edison as his fame spread globally. They facilitated Edison's ready access to the press throughout the country. Other reporters whom Edison had known during his years as an itinerant telegrapher, particularly Edwin Fox (*New York Herald*) and Thomas Maguire (*Boston Globe*), resurfaced during the phonograph launch offering him additional press contacts as well as advice on how to handle the media.

Although most of Edison's networking endeavors occurred in one-on-one and small group sessions, he also availed himself of high-level opportunities to display his inventions and meet new contacts at trade shows in the United States and in

Europe. Edison's first major trade-show opportunity was the Centennial Exhibition in May 1876, in Philadelphia. The Philadelphia show featured his automatic telegraph displayed at the Atlantic & Pacific Telegraph Company booth, while his other printing telegraph inventions and duplex machines were presented at the Western Union booth. Edison had sold patent rights for these inventions to the companies who proudly exhibited them at the show. And, he also rented his own booth to display his electric pen, for which he'd maintained sole rights.

Edison's electric pen, noted by the Exhibition judges for its "exquisite ingenuity and . . . usefulness," won several major awards at the show. Edison's successes at the Centennial Exhibition impressed many luminaries, including famed British physicist Sir William Thomson (Lord Kelvin). Lord Kelvin made special arrangements to tour Edison's Menlo Park laboratory in July, a visit that created an important professional relationship between the two men, linking Edison to the influential British scientific community early in his career.

Edison reaped the benefits of displaying his work at the Philadelphia trade show, and he and his teams became ardent participants in these types of events around the world. In early 1878, he hired eighty men to staff his booth at the Universal Exposition in Paris, a huge international trade show, which ran from May through October. His team amazed the huge crowds with their demonstrations of the phonograph. Edison traveled to Paris for a few weeks of the show, receiving the Grand Prize Medal for "Inventor of the Age." Edison's patent agent, Polish businessman Theodore Puskas—whom Edison had engaged years earlier to negotiate and oversee his international patent interests—set up numerous networking appointments for Edison, primarily with major financiers who later became important in expanding his European lighting business in the 1880s.

In the spring of 1881, Edison sent Charles Batchelor to Paris to set up a mammoth trade show booth at the International Electrical Exhibition. The booth included an actual working generator weighing more than thirty tons; it was intended to form the core of an actual central power station that would remain on site in Paris for several years. Edison spared no expense in his determination to demonstrate the superiority of his system versus any other at the Exhibition.

Edison's booth served as a pivotal networking center. Although Edison himself did not attend the International Electrical Exhibition, he arranged for Batchelor to receive visits from key representatives of major European companies, including Siemens & Halske, leading to a number of productive partnerships.

Years later, when Edison arrived in Europe for the 1889 Paris Exposition, he was revered as a legend in his own time. The 1889 Paris Exposition was a global scientific gathering whose gargantuan scale rivaled that of large trade events today. In Edison's display room—occupying nearly an acre of space—"every day some 30,000 people heard twenty-five phonographs talking in dozens of languages." During his eight-week stay, Edison met privately with Alexandre-Gustave Eiffel, architect of the Eiffel Tower, and with French physiologist Étienne-Jules Marey, a meeting that spurred Edison's development of motion picture technology. Marey's work using photographs to capture animals in motion served as a key inspiration for Edison. Prior to his departure, Edison also met with Werner Siemens in Germany, where he was also invited to attend a meeting of the German Association for the Advancement of Science in Heidelberg. Through these key networking contacts and countless others, Edison further strengthened his position within the business and scientific communities of Europe.

Edison loved trains. In this photo taken in 1918, the "engineer" is Henry Ford, the "fireman" is Harvey Firestone, and Edison rides the cowcatcher.

Edison benefited from his networking efforts on many levels, but he also practiced the core success principle articulated by Napoleon Hill: "You give before you get." Edison sought always to help and encourage others, including the many young entrepreneurs and business people seeking his counsel.

Henry Ford was one of them. A chief engineer at the Detroit Edison Company, Ford was attending an annual meeting of Edison company engineers in New York in August 1896 when Alexander Dow identified Ford to Edison as "a young fellow who has made a gas car." Edison engaged Ford in a conversation that ended with Edison banging his fist on the table, exhorting: "You have it! Keep at it!"

Years later, Ford told Edison he was the very first individual who encouraged him to continue with his "crazy ideas" for a gasoline-powered car.

Creating Innovation Literacy: Become a Master Networker

"The Chinese call it 'guanxi.' And they believe it to be the very basis for successful business leadership. In other words, it's not just what you do, it's largely who you know," notes Professor James Clawson of the Darden Graduate School of Business. As Edison knew, however brilliant you might be, you only get results by working with others. Becoming a master networker is an essential element for creating *master-mind collaboration*, both for your personal success and the realization of your organization's innovation initiatives.

As National Inventors Hall of Fame inductee Dr. Donald Keck comments on the importance of networking, "My admonition to my scientists was, 'Get out around the world—there's no geographic boundary to a good idea. Go to a conference.'" A few years ago, at a national conference for three thousand librarians, we watched Susan RoAne, author of *The Secrets of Savvy Networking,* as she taught the group how to become master networkers. We made a point of becoming part of Susan's network and the result is that we were able to consult with her to offer you these thoughts on applying this important element of Edison's *master-mind collaboration.*

- *Assess and maintain your current network.* You may be surprised at the extent of the network you already have. Master networkers like Edison know whom they know. Now, of course, you can do this via a contact management program or, like Edison did, by listing your contacts in a special notebook. RoAne suggests implementing "a plan for regularly 'touching base' to solidify relationships with your network. Make a couple of calls and a few emails each day to people in your network. These communications should focus on something that might interest or benefit the recipient."

- *Focus on diversity.* It's easy to have a network of people just like ourselves, but sameness doesn't serve your evolving master-mind collaboration efforts. Like Edison, cultivate connections with people of different generations, backgrounds, interests, specialties, and geographies. As RoAne points out, "You never know the potential benefits that come from meeting someone who at first may seem unimportant."

- *Target "key influencers."* Edison targeted, through careful research, the experts and key influencers in his special areas of interest. You can do the same. Make a list or mind map of key people you want to meet, then find a way to make it happen.

- *Stay visible.* Edison attended major trade shows and industry meetings at a time when traveling to such events wasn't easy. Master networkers err on the side of accepting invitations to important conferences, meetings, programs and events, even when they would prefer to stay home.

- *Work the room.* It is easier to talk to people you already know, so it takes a conscious effort to reach out and meet new people. The challenge, as RoAne emphasizes, is that 85 percent of people consider themselves shy. And, as she quipped on the day we met her, "The percentage is probably higher for librarians." RoAne suggests that you can overcome reluctance to engage in conversation with new people and free yourself from worry about awkward silences by "having a simple, planned seven- to nine-second self introduction." She adds, "Like Edison, keep informed, and you'll always have topics for conversation."

- *Act like a host.* Edison was often a host in his trade-show booths, and taught his staff to be welcoming hosts as well. RoAne emphasizes the importance of

ORGANIC NETWORKS

Professor James Clawson is a valuable member of our master-mind collaboration network. He emphasizes the role of organic networks in the process of organizational innovation. He comments, "Organizational theorists have known for decades that the "real" organization, or the informal organization, is not necessarily what's on the organization chart. This organic or shadow organization is based on the network of relationships that people throughout an organization have developed. This network plays a critical role in translating ideas into innovations."

Clawson adds, "Edison was a master networker and he loved to express ideas visually. In *The Hidden Power of Social Networks*, Rob Cross, a professor at the University of Virginia, introduces a networking tool that would make Edison smile. Cross has created a simple but powerful way of collecting online data from organizational members, then displaying that information in dramatic network diagrams showing clearly the strengths and weaknesses the informal organizations develop behind the formal organization chart. Cross has developed a networking roundtable with a blue chip list of member companies in a remarkably short time. The participant companies, the model he's using, and a description of his process are now available online."

acting like a host at any event. In other words, as you meet new people, introduce them to others. When you provide introductions, as a gracious host would in their own home, you make other attendees feel more comfortable, engaged, and connected to you.

- *Put others first.* Edison offered help to others through introductions, ideas, references, information, and encouragement. He expanded his network by sharing it with others. As you work the room, think about what each person you meet might want or need that you can provide. As Napoleon Hill frames it, "The best way to sell yourself to others is first to sell the others to yourself."

- *Follow up.* Follow-up establishes credibility and distinguishes you from the many people who attend events, hand out cards, promise to stay in touch, then don't. When you attend an event and walk away with eight business cards, for example, review them; within two days, send a brief email or handwritten note to acknowledge the connection. It's as simple as writing, "Glad we met. I was fascinated by your description of the patent process."

- *Keep it personal.* Although e-mail, Skype, and cell phones make it much easier to stay in touch with people all over the world, it's important to remember there's no substitute for personal, face-to-face connection.

Edison's approach to *master-mind collaboration* allowed his teams to be exceptionally productive in generating, developing, and testing his innovations. Edison always understood, however, that the ultimate purpose of all their efforts was to create exceptional value for customers. His early experience with his technically brilliant but ultimately useless vote-recording machine taught him a lesson he would never forget: Inventions must be useful for real people. Edison never took his eye off the market. And he pioneered an approach to creating customer value that is even more relevant in our hypercompetitive global marketplace. We call it *super-value creation*, the fifth competency for *Innovating Like Edison*.

Chapter Seven

COMPETENCY #5— SUPER-VALUE CREATION

Carlson and Wilmot define innovation as "the process of creating and delivering new customer value in the marketplace." Edison's philosophy of value was "bringing out the secrets of nature and applying them for the happiness of man." We call his approach *super-value creation*. Why "super"? Because it suggests creating value above and beyond your competitors. It is the ultimate innovation competency. Edison knew of "no better service to render during the short time we are in this world."

Once he had gathered information about openings in the market and the needs of the consumer, Edison analyzed how his observations meshed with what his laboratories could deliver—or could learn to deliver. He then calculated how much it would cost to go after the market—or markets—he had in mind, creating an innovation plan including commercialization options. And finally, he placed the finishing touch on his products: the mystique of the Edison brand name.

Edison was a master at anticipating trends and spotting gaps in the marketplace. His approach used both analytical and intuitive tools to help determine market size and the best target audience. In Competency 5, we'll show you how to use them for your audience. Thomas Edison drew customers to his products with sophisticated branding techniques plus a wide array of media and communication tools. With Edison as your guide, you will learn how to design a business model that is best suited for your ideas, or for your organization and its innovation endeavors.

The Elements of Super-value Creation are:

21. Link Market Trends with Core Strengths
22. Tune In to Your Target Audience

23. Apply the Right Business Model
24. Understand Scale-up Effects
25. Create an Unforgettable Market-moving Brand

ELEMENT 21: LINK MARKETPLACE TRENDS WITH CORE STRENGTHS

Edison began all his innovation efforts by asking himself questions such as: "What needs do people have that I can fulfill? What trend or trends are present here? What opportunities do they present? What are the current gaps in the marketplace? What is the insight that can lead me to create greater value in this segment? How can I leverage what I know about this category, or industry that makes sense for my laboratory, and my brand name? How can I test the efficacy of my idea?"

Edison looked for trends through his reading and networking. He would note places where he thought something was missing, or areas where he felt the quality, efficiency, or technology might be improved. He then linked the gaps and trends, and started generating insights about them in his notebook. Using hypotheses he developed about each linked trend and gap, he would work on ways of fleshing these out further himself, or assign a team to the task. Whatever the project, Edison focused on how he could deliver the greatest possible value to the marketplace. Of course, Edison did more than just develop innovations by following trends, he ultimately *created* trends through his innovations.

Edison in his West Orange office (1912) examining a reel of film for his new kinetoscope, a movie projector designed for home use.

Because Edison liked everything to work "with the least amount of friction," he had a keen eye for inefficiencies, including unnecessary uses of labor, extra moving parts in equipment design, and cumbersome technical processes. Many of his early inventions focused on ways of making telegraph machinery operate more efficiently. Efficient equipment reduced manufacturing and maintenance costs and provided greater ease of use for his customers. Edison saw opportunities for improvement at every turn. His major challenge was to organize them in a fashion that linked his core strengths with the creation of maximum value for his customers.

Making the leap from spotting an area for improvement or a marketplace gap, to implementing an innovation within the parameters of your brand and core competencies is extremely challenging. Analysis of Edison's work indicates he had an intentional, five-step process for linking trends to his company's core strengths, which we share with you here.

Edison's five-step process flows as follows:

1. **Trend**: Identify trends in the marketplace, observing where and how needs are shifting.

2. **Gap**: Determine if there are any gaps in your field created by these changing needs. (You can of course look at gaps for other industries besides your own.)

3. **Insight**: Identify the "core insight" (the "aha!") or need at the heart of each gap.

4. **Linkage**: Link your insights to your capabilities as a company and your strengths and weaknesses in relation to competitors. If you require new skills to implement a linkage item, assess your ability to develop those skills in a cost-effective manner.

5. **Hypothesis**: The hypothesis is an "if then" proposal for a practical experiment about the goal you'd like to fulfill based on the "linkage" you've created.

Edison applied this five-step process throughout his career. To give you a taste of his approach, we present three examples, written as Edison might have noted them, demonstrating how he organized his thinking about generating innovations.

Case #1: Inventing high-quality equipment related to stock reporting and financial services.

1868 Trend: I read in the newspapers that the rebuilding of the American South after the Civil War has sent commodities prices rising, since demand for lumber, brick, stone, and other building materials has increased and supplies are strained. The reporting of these prices—as well as the price of gold and precious metals—is eagerly sought by American businesses at an increasing rate.

Gap: I've noted that few efficient machines and few services exist to rapidly—and reliably—report commodities prices, precious metals prices, and stock prices to businesses around the country.

Insight: I think I can apply the principles of telegraphy to make a machine that reports prices.

Linkage: I can combine my talents for telegraph equipment assembly and design to develop price-reporting machines that operate rapidly and efficiently.

Hypothesis: If I can build an efficient, reliable telegraph-like machine that can help businesses report commodities and metals prices all over the country, then I will make a significant profit.

Result: Edison's stock-reporting machine launched his career as a successful independent inventor and innovator.

Case #2: Development of a system that will create multiple copies of a single document.

1872 Trend: I've noticed that people working in offices where they sell insurance, or where they do accounting work, all now seem to be using more paper. They stay very busy making documents for lots of people to see and use.

Gap: There is nothing that allows one person writing a document to easily and quickly make more than one copy of it.

Insight: I think a battery could be used to make a piece of equipment that creates a stencil that can make lots of copies of one document. I can find a way to reproduce the indentations needed for a stencil by making some kind of special pen.

Linkage: I know a lot about making chemically treated paper. I know how to make paraffin interact with paper, leaving a smooth surface. I also know how to make neat, precise perforations in paper. I understand a lot about how telegraph equipment indents paper, and how small motors and batteries can be used to create a mechanical system.

Hypothesis: If I can create a portable system that can be used in offices to reproduce copies of documents, then I will launch a new market.

Result: Edison invents the electric pen in 1875, successfully marketing it in the United States and Europe, and later licensing it to the A. B. Dick Company. A. B. Dick expands distribution of the system, which ultimately becomes known as the Edison Mimeograph Machine.

Case #3: Examining how sound waves impact different kinds of materials.

1875 Trend: I notice that an increasing number of scientific experiments are being conducted in the United States and Europe on acoustics, examining ways of making sound travel more effectively through different media, and determining how sound travels through the human ear.

Gap: I see only a few places where a deeper understanding of sound and acoustics science is being applied to telegraphy. Bell has beaten me to the telephone. But there isn't a telephone that records a sound message when no one is there to answer.

Insights: I note that sound travels in waves, just like telegraph communications do. Sound waves also make permanent indentations on various media.

Linkage: I can conduct experiments that combine diverse uses of acoustical equipment I am familiar with through my reading of German acoustics expert Hermann von Helmholtz, including reeds, tuning forks, and metal strips.

Hypothesis: If I can create equipment to capture the rapid rate of vibration created by sound, then I have a recording telephone.

Result: Instead of the first voice mail system, Edison's work with acoustics and his knowledge of telegraphy lead to the invention of the first tinfoil phonograph in 1877.

Edison's five-step approach drove development of his successful products and services, and set the stage for modern innovation. Examples of those who have followed in his footsteps were abundant in a recent issue of *TIME* magazine featuring the best inventions of the year.

The article described a wide range of exciting new products, including the "Hug Shirt," a high-tech garment that uses wireless technology to give the wearer a virtual hug. The shirt's circuits are programmed to replicate the pressure, heartbeat rhythm, warmth, and length of your designated cuddle buddies (www.cutecircuit.com). The article also describes a drip-free umbrella (www.proidee.co.uk), a scorch-free iron (www.oliso.com), a robotic sommelier (www.necst.co.jp), and a pain-free way to dispatch crustaceans (www.crustastun.com).

All these products are aimed at filling gaps in the marketplace, addressing customer needs ranging from the desire for affection, to the need for guilt-free lobster consumption. But, the number one invention of the year in 2006, according to *Time*, was YouTube. Steve Chen and Chad Hurley started YouTube as a personal video-sharing service in the winter of 2005. Hurley and Chen focused on a trend, the increasing popularity of Internet networking sites, and they identified a gap: a lack of video networking online. They came to an "aha!" core insight, realizing that they could assemble the technology to allow people to upload and easily view their videos. Hurley and Chen linked their insight to the skills they possessed, then hypothesized that if they could get a video-sharing service working then they would attract a significant following, opening possibilities for different commercialization strategies. The result: a consumer media company with people watching more than 100 million videos on their site daily. YouTube was purchased by Google for $1.65 billion in stock.

Creating Innovation Literacy: Link Marketplace Trends with Core Strengths

Market shifts are driven by technological innovation, shifting demographics, fluctuations in the supply and demand of commodities, climatic phenomena, religious movements, and political developments such as interest rates and tax policy changes or the opening of formerly closed economies like India and China. If people didn't age, move around, or innovate, if new infectious diseases didn't occur, if the weather was constant and governments were all stable, then things would probably stay pretty much the same—and trends wouldn't be important. Of course, things don't stay the same, and the pace of change is accelerating dramatically, hence the importance of anticipating trends.

Here are a few examples of the directions in which we think things are moving globally:

Increasing Speed: Computers are getting faster, the ability to exchange information is accelerating, the pace of change is almost dizzying, and so is the demand for even more speed. Consumers want immediate customization, quicker ordering, and speedier delivery of all the products and services they consume.

Shrinking Size: In a prophetic speech in 1959, Nobel Prize winner Richard Feynman said, "The principles of physics, as far as I can see, do not speak against the possibility of maneuvering things atom by atom." Nanotechnology is the rapidly emerging science of the miniscule. Computing power once available only through massive supercomputers will soon be harnessed on a single computer chip through emerging technologies like the Cell Broadband Engine, jointly developed by Sony, Toshiba, and IBM. Mini-supercomputers will allow major new capabilities to be embedded in medical devices, television sets, and computer games, as well as virtual reality consoles.

Smarter Objects: Last year, the world produced more computer chips than grains of rice. Some microprocessors are now so small and inexpensive they've become disposable. Expanded uses for RFID (Radio Frequency Identification) tags and other chip technologies will mean everyday items can become "smart," creating huge new markets.

Greater Transparency: Access to timely information continues to accelerate globally. The ubiquity of camera phones and video surveillance combined with the emergence of YouTube and other image-sharing sites means there's an increasing likelihood that, whatever you do, it may be recorded and shared with the world.

More Collaboration and Co-creation: Collaborative relationships and "open source" communities, enabled by the Internet, will further accelerate the creation of new intellectual capital. Consumers will have greater opportunities to contribute to the customization of the products and services they purchase.

Reducing Carbon-based Emissions: Global awareness of the importance of reducing carbon-based emissions is growing. Momentum is increasing to find viable technological solutions to global warming.

Intensifying Competition: The continuing development of high-tech/low-cost economies in India, China, and elsewhere combined with the proliferation of fast, inexpensive communication systems will intensify competition for products and services globally.

In the United States, a few obvious, broad-based trends include:

Demographic Shifts: An aging, affluent population will reshape the meaning of retirement, and strain social service systems for decades. Continuing expansion of the Hispanic population will raise the need for more Spanish-language media and product usage information.

More Visual Communication: An increasingly diverse population will drive more reliance on images for communication rather than words.

Expanding Commercial Outreach by "Nonprofits": Increasing competition for nonprofit dollars and advances in marketing-communication technology will drive expanded branding efforts by educational, charitable, health care, and other nonprofit institutions.

Increasing Desire for Privacy and Security: The ubiquity of security cameras and associated monitoring devices will drive expanded desire for privacy, even as we seek greater protection from terrorism and crime.

It's useful to begin your own efforts to link marketplace trends with core strengths by considering this kind of "macro" perspective on trends. The trends listed are offered to stimulate your thinking: Do you agree that these trends are significant? If so, what are their implications for your life and work? What other trends can you identify, and what are their implications? Edison enjoyed this kind of exploration and speculation on a purely intellectual basis, but he also used his awareness of trends to drive his practical innovation and *super-value creation* strategies.

Just like the folks at YouTube, you can begin applying Edison's five-step system immediately. Identify trends in areas that interest you and look for gaps. Apply your *kaleidoscopic thinking* skills to generate insights, and create linkages that lead to hypotheses or proposals for action.

In addition to reading and thinking about macrotrends, increase your awareness of trends in your field by doing what Edison did: Read a wide range of journals, periodicals, newspapers, and books in addition to technical journals in your field; also read general-interest publications, and remember to include a few publications originating in other parts of the world. Make a habit of reading a few things every week you wouldn't normally read. Seek out quirky, offbeat perspectives. Of course, trends manifest faster now than they did in Edison's time, but you can use Internet research to keep pace. Nielsen BuzzMetrics is one of a growing number of firms specializing in mapping trends on the Internet through strategic blog analysis.

As you identify trends that interest you most, begin to look for gaps. Peter Lynch, the legendary manager of Fidelity's Magellan fund, made billions of dollars for his investors through his ability to read trends and spot gaps. His most famous investment principle is, "Invest in what you know." Lynch emphasizes the importance of "local knowledge." In other words, pay attention to trends and tendencies in areas familiar to you. In his classic, *One Up on Wall Street*, Lynch describes how he discovered many of his best investments by noticing gaps in the marketplace in the course of his everyday life.

You can look for gaps in lots of places: quality, technology, process, efficiency, user-friendliness, and pricing. A participant at a recent innovation seminar shared the following observations about some of the gaps she observed:

Quality: "The carbonated cola I just opened this morning and resealed has already lost its fizz after only a few hours."

Technology: "My laptop computer doesn't understand input from my digital camera."

Process: "It's so hard to understand what I'm supposed to do just to put air in my tires at a gas station. I can't figure out the gauge for reading the tire pressure."

Efficiency: "My medical records are scattered across several different hospitals in different states, and I somehow have to find a way to gather them up before my knee surgery next month."

User-friendliness: "When I call my cable television company, I have to go through four levels of automated menus before I can either get my question answered or speak to a real human being."

Pricing: "It's so expensive to ship furniture purchased online."

This "gap awareness" may not seem earth shattering at first, but cultivating it in everyday life offers a simple, powerful, and quick way to strengthen your ability to *Innovate Like Edison*. Of course, translating your gap awareness into a successful innovation requires linking it to your core competency. Here's an example from a recent innovation seminar.

A four-person management team from a successful midwestern design consultancy came to the seminar to generate new approaches for a client assignment they'd recently received. The team had run into several blind alleys and they were stumped about how to begin their new project.

The design firm's assignment involved identifying a new delivery system—and a new branding concept—for their client's teeth-whitening system. The team's initial research made it clear that aging baby boomers were increasingly whitening their teeth, as well as using antiwrinkle regimens and hair colorants to maintain a youthful appearance.

The team knew that several major toothpaste brands were already offering gels and whitening "strips" that could be worn for a few weeks.

Complicating the matter further, many dentists nationwide already featured complete one-visit teeth whitening systems ranging from $300 to $600 each—a price their client was willing to meet. However, the client did not want to offer its new line through dental offices.

The design group framed their challenges: How can our client's whitening approach be distinguished in the marketplace? How can it be profitably optimized to create maximum value for customers? What trends connect our client's product to existing momentum in the marketplace?

To begin their analysis, at the top of a large piece of paper, the group wrote: "Creating Super-value Teeth Whitening."

Next, they listed across the top of their paper a handful of competitors who were already offering the kinds of products or services their client was considering. The design group listed: Rembrandt (one of the first and most successful teeth-whitening systems offered in mass distribution), Crest Whitening Strips, Aquafresh Whitetrays, and in-office dentist treatments.

Then, the team held a "green hat" Ideaphoria session for ten minutes, generating a list of the factors they felt most influenced the perceived value of whitening products for consumers, that might also drive brand success. They generated a list of almost thirty factors, which they narrowed to eight, with ten being the maximum recommended. Their list, noted down the left side of their paper, included factors like "Easy to use," "Pain-free," "Easy to access," "Fits with my lifestyle," and "Easy to pay for," along with several others.

At the bottom and top of the chart, they created a rating scale from 1 to 5, with 1 being "Does not meet this criterion" and 5 being "Fully meets this criterion." They chose different line patterns for each competing product or service, and made a legend at the top of the page showing which pattern represented which item. Competitor by competitor, the design group worked through each factor, giving it a rating, and "mapping out" the estimated performance for each product with a marker. This led to a series of zigzag lines across their chart, similar to the one shown in the illustration on page 182.

The team then looked for the "white space" on their chart, representing opportunity gaps where no one was presently providing value to the consumer. Using a different colored marker, the group drew a new line through the white spaces on the chart. This "gap line" revealed areas where opportunities to provide super-value in the teeth-whitening market still existed.

Looking at their "gap line," the team considered how it fit with trends they knew about. For example, they commented that many baby boomers had some kind of workout routine—something simple like walking or belonging to a health club. Thinking that health might be a macro-trend that related to the factor "Fits with my lifestyle," the team placed an asterisk (*) next to the gap line and noted down the words "Health Clubs." They also felt there was a trend related to the ease of paying for transactions with a debit or credit card. The team added an asterisk (*) near their gap line by the factor "Easy to pay for."

After focusing for several minutes on the gap line plus the two trend asterisks, the team had a major insight: Why not offer the whitening system as part of a health club membership? Or even a country club membership? Customers could apply the product to their teeth after working out (or playing golf, swimming),

CREATING SUPER–VALUE: TEETH WHITENING

Legend: Rembrandt Whitening Toothpaste   ....
           Crest Whitening Strips   – – – – – – – – – – –
           Aquafresh Whitetrays   – · – · – · – · – · – ·
           In-office dentist treatment   + + + + + + +

           "Gap Line"   O O O O O O O O O O O O O O O O O O

PRIMARY FACTORS	Does not meet this criterion			Fully meets this criterion
	1	2	3	4	5

1. Easy to use

2. Pain-free

3. Easy to access

4. Fits with my lifestyle   Trend: *Health clubs

5. Easy to pay for   Trend: *Credit or Debit card

6. Fits within my budget

7. Will generate lasting results

8. Makes me feel like I'm doing something healthy for myself

1   2   3   4   5

Visually mapping markets and their characteristics reveals "white space" gaps for new product or service ideas.

then stand in front of a special activation light—just like the ones used in dentist's offices for whitening treatments—in the locker room, beaming it right onto their teeth. The cost and length of dosage could be monitored by a Smart Machine right on the premises, with a phone nearby for customer service if needed. Either a card-swipe system or a personal identification number (PIN) could be used for payment, or members could sign up for monthly plans.

They knew their client had a core competency in teeth-whitening technology that would allow the product to be offered in a variety of forms. What they now realized, however, is that their client would either need to develop a new core competence involving distribution, or partner with another firm to accomplish the distribution component of this idea.

The team began buzzing excitedly, creating a series of "If . . . then . . ." hypothesis statements for their project, as well as a group of experiments they could conduct to further their thinking and test the efficacy of their "aha!"

By selecting relevant competitors, listing key value and success factors for their industry, and rating the level of perceived delivery for each competitor by factor, they "mapped the gaps" and then combined their findings with trends, yielding significant insight for their client. They left the conference ready to swing into action with their first experiment: identifying how they could start generating super-value—and whiter teeth-for baby boomers.

Desiree Gruber, founder and president of Full Picture, a public relations firm in New York City, describes how she linked marketplace trends with gaps to facilitate the emergence of a different kind of supermodel:

> When I first met Heidi Klum in 1996, she was working with the Elite modeling agency in New York, yet still relatively unknown in America. I knew she faced challenges because—despite her beauty—she did not fit the mold of the top U.S. fashion model . . . Most supermodels were extremely thin, haughty and aloof, but Heidi was full-figured, open and accessible. You can imagine how refreshing this was. The way to build Heidi's career was to focus upon what distinguished her from the rest rather than suppress it.
>
> It was apparent to me that Heidi filled a gap in the fashion world. Her accessibility and naturalness made it easier for people to feel an affinity for her. One delightful example was when she showed off her yodeling skills on the David Letterman show. It didn't occur to her that some might consider it

risky for an up-and-coming model to reveal anything but total glamour. Yodeling was simply a skill she had, so she shared it.

As I watched Heidi grow as an individual, it seemed only natural that the variety and scope of her work would grow too. So we proposed a reality-competition television show built around her strengths: style, beauty—but also acumen and sensitivity. The result was *Project Runway*, the highest-rated show in the Bravo cable channel's history, and a pop culture phenomenon.

Because we emphasized what makes her different, and created spaces where these differences could shine, Heidi was able to realize her true potential. We saw a gap in the fashion market for a "real" model who was genuine and accessible, and linked Heidi's "core competencies" to this powerful new image.

ELEMENT 22: TUNE IN TO YOUR TARGET AUDIENCE

> Anything that won't sell, I don't want to invent. Its sale is proof of utility, and utility is success.
> —*Thomas Edison*

In a recent online review of the Pearl Street Diner in Lower Manhattan, a happy customer wrote: "Being new to the Financial District, I stumbled across the Pearl Street Diner while taking a stroll during my lunch break only to find the best place for burgers, meatloaf, salads and my favorite jumbo stuffed shells . . . hmmm. Their gyros are also good with real thick and rich sauce on the side. I recommend anyone in the area to stop by and try their food. I always get speedy service with a smile."

The Pearl Street Diner knows what their customers want: delicious food and friendly, efficient service at a competitive price. Of course, the diner didn't exist when Edison walked these same streets 130 years ago to find out what his customers wanted and needed. The neighborhood is quite different now, but tuning in to your target audience is as important as ever whether you're dishing burgers or lighting the world.

Curt Carlson and Bill Wilmot of SRI International have developed a culture dedicated to creating customer value through practical innovation. They pay tribute to America's pioneer of this approach:

Thomas Edison was a master at knowing when to attack a problem, and he created one marvelous innovation after another . . . The objective is, as Edison knew, finding solutions to important customer and market needs where all the pieces can come together. The electric light bulb was an invention. The creation of the electric light bulb with a practical electrical distribution system that could economically deliver power to customers was an innovation—one of the most important in the history of mankind.

As Carlson and Wilmot emphasize, Edison's development of a system of electrical power represents one of the most ambitious and significant strategic innovations of all time. It was successful because Edison focused his inventive prowess to meet important customer needs in an economically viable way.

Edison's understanding of the importance of tuning in to his target audience began early in his career with the failure of his vote-recording machine. The machine worked perfectly, but it didn't sell because legislators didn't actually want timely and accurate vote counting. This precipitated an epiphany for the young inventor: Edison realized that for an invention to become an innovation, and to be more than just a good idea, it had to meet customer needs. Edison resolved to focus all his efforts from that day forward on understanding and meeting the needs of his customers. As he noted, "Anything that won't sell, I don't want to invent. Its sale is proof of utility, and utility is success." Edison's early epiphany around the central importance of customer needs guided his career.

Edison's creation of the electric pen presented him with his next significant lesson about providing what customers wanted. Edison knew there was a growing demand in the business world for ways to duplicate documents. He set out to invent a copying machine that could make a hundred copies. As Paul Israel notes, "He had high hopes of finding a ready market among merchants, lawyers, insurance companies, and other firms that 'seem to have a great deal of reduplication.'"

Edison's invention transferred ink from a master paper stencil to a small printing press. Passing clean sheets of paper across the inked stencil produced multiple copies. Edison designed a vibrating metal stylus connected by fine wires to a small battery-powered motor, allowing the pen to make the fine holes in the stencil. Each unit consisted of the pen, battery-driven motor, special ink, and a printing press housed in a decorative cast-iron cabinet. The Edison Autographic Press & Electric Pen retailed for $30 (the equivalent of $505 today).

Edison hired independent sales agents to sell his electric pen. As he anticipated, there was great demand for his product. But one day, one of the sales agents, an ex-telegraph operator named Mullarkey, informed Edison of several difficulties

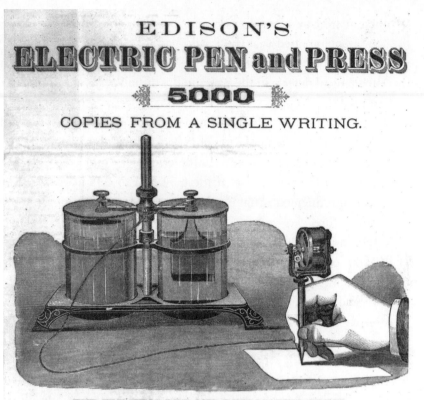
This advertisement for the electric pen highlights its acceptance by customers around the world. Consistent with the focus on practical value associated with the Edison brand name, the ad lists more than fifty uses for the pen, as well as how documents created with it could be sent inexpensively through the mail.

encountered by users. In addition to complaints about the noise of the motor, Mullarkey reported that those who had to "carry the outfit from office to office, found the box is infernally heavy." There were also numerous complaints about "mechanical defects in the pen and the difficulty of taking care of the battery that powered it."

Edison and his right-hand man, Charles Batchelor, took Mullarkey's reports seriously. And, instead of just going back to the lab to work on the technical problems, Edison sent a team out to the customers' offices to observe how they used the pen in their work. From these visits, Edison's team "identified fifteen mechanical refinements that not only improved the pen but also made it cheaper to manufacture."

Following these improvements, the electric pen became a major success and Edison soon had an expanding network of sales agents. As William Orton of Western Union, an investor in the product, noted, "Indeed, it has been difficult to supply the demand."

After growing the business based on a more thorough understanding of how consumers used his electric pen, Edison was able to sell it for a handsome profit and move forward with other projects. Edison invested all his profits in a new venture: a system for electric lighting. He also invested his now-refined understanding of the importance of tuning in to his target audience into all of his endeavors. Edison's experience with the electric pen led him to not only target areas of important customer demand, but also to better understand the application of his products from the customer's perspective. And, he realized it was better to develop this understanding before releasing his products.

Before introducing his system for electric lighting, Edison conducted in-depth consumer research. Edison's teams canvassed every apartment and office in a six-block area in lower Manhattan surrounding Pearl Street. Edison had learned from his electric pen experience that it was critical to observe consumers in the environments where his products would actually be used.

Edison taught his research teams to observe the patterns of his future customers' usage of gas lanterns, stoves, and heaters. He gained invaluable information about the strengths and weaknesses of his competition. All the data was compiled and then analyzed by Francis Upton. Edison used what he learned from these studies to fine-tune his invention work; and, with Upton's help, he completed the laborious process of forecasting total market demand for the area. The involvement of his staff on these missions was critical in attuning them to consumers' needs as well.

Edison was a pioneer of what we now call "ethnographic research," a segment of market research studying patterns of behavior and circumstances of use for a given product or service. The chief advantage of this form of research lies in the *nuances and detail* available to the observer. Often, these nuances are lost when other forms of research are undertaken.

After the success of his Pearl Street experiment, Edison applied his audience-targeting strategies to the cities and towns that seemed to be the best candidates for new power stations. He engaged and trained special agents to do his market research. His agents were assigned the task of identifying neighborhoods of prospective residential and business customers with high profit potential. After finding appropriate districts of approximately one square mile, Edison's canvassers would go street-by-street conducting interviews with each possible customer. This data was essential to the development of budgets and plans for the construction of each power station.

Edison spent $10,200 of his own money ($193,000 today) doing consumer research for eighty communities, of which he won contracts for twelve. Of course, as a result of these successes, it wasn't long before the whole world wanted the benefit of Edison's work.

Edison's focus on targeting his customer's needs became part of his organizational culture at both Menlo Park and West Orange. One of the busiest rooms in Edison's West Orange laboratory operation was the first-floor space devoted to reading customer correspondence. Edison devoted time to reading portions of this flood of correspondence:

The serious and important part of the mail, some personal and some business, occupies the attention of several men; all such letters finding their way promptly into the proper channels, often with a pithy endorsement by Edison scribbled in the margin . . . The amazing thing about it all is that this flood of miscellaneous letters flows on in one steady, uninterrupted stream, year in and year out . . .

It was through written customer correspondence that Edison received word of a slight quality defect in his storage battery. After a five-year development period,

Cable Address "Edisons NewYork"

From the Laboratory of Thomas A. Edison,

Orange, N.J. Aug 14 1918.

Dear Canty

In reply to your question, let me say that I was the first person to speak into the first phonograph. The first words spoken by me into the original model and that were reproduced, were "Mary had a little lamb" and the other three lines of that verse.

Yours sincerely

Thos A Edison

Edison's distinctive handwriting is evident in this letter he wrote at age seventy-one in response to a question about the phonograph.

and the manufacture and sale of thousands of units, Edison learned through a handful of letters from customers that there was an occasional minor seepage along the outer seams of the battery. He responded, despite the objections of his laboratory and manufacturing staff, by closing his storage battery manufacturing line and discontinuing sales. Five years and more than fifty thousand experiments later, he released an improved storage battery. It became an extremely profitable item.

In his laboratories as well as his manufacturing facilities, Edison drove the process of continuous improvement through an insistence on understanding how consumers perceived his products and services. Edison's ability to tune in to his target audience led him to forge new insights about how to create super-value through his innovations. He pioneered the process of developing and testing prototype products with his target audience, applying the knowledge gained to further improvements prior to launch. After launch, he maintained a connection to consumers by reading their letters, and responding to their displeasure or delight.

Creating Innovation Literacy: Tune In to Your Target Audience

At a seminar for some of the top engineers of one of the world's leading chemical products companies, the group was shown a video of presentations they had given to their counterparts in marketing. The marketing folks had commented they had trouble understanding the presentations, and it was easy to empathize with them. One presentation after another focused on incomprehensible technical details. One engineer referred repeatedly to the complexities of the molecular structure of the compound he was investigating. After watching the video, the group sat in silence for a minute and then their leader stood up and shared an insight. As he put it, "Oh my, no wonder we have trouble communicating; we are guilty of molecule-fondling!"

"Molecule-fondling" is a marvelous term for the tendency of technically oriented people to focus on their research and forget about their audience. And, the more brilliant and passionate the researcher is, the easier it becomes to lose touch with end-users.

Rich Sheridan, CEO of Menlo Innovations, describes his own experience of overcoming molecule-fondling and learning to *Innovate Like Edison*:

> For over twenty years, I have been working in, with, and for, software product companies . . . I've made many mistakes. I've built products that customers simply didn't want. I've wasted valuable dollars building features users couldn't use.
>
> Fortunately, I've learned . . . I've learned there is a way to build software so that customers are happy to pay for the end result—without a lot of selling. The trick, and this is really hard to say for a recovering computer programmer, is to not focus on technology."

Sheridan explains,

> Before this change, when I was a computer programmer, it was obvious to me that many of my problems were not, in fact, technical. After all, I created "great" products, with cool new technology. Why couldn't the customers see how great it was? It was a hard thing to realize that the problem was not with my customers or users. The problem, in fact, was with me.
>
> I found that it's really impossible to know if you are building a system that helps or hurts your customers unless you have a profound understanding of how they really do their jobs.

Sheridan and his team discovered that the best approach to tune in to their target audience came not from computer science but from anthropology. They studied the techniques and methods of anthropology and applied them to understanding the system design process from an end-user perspective. Sheridan calls his approach "High-Tech Anthropology." He describes the benefit: "Amazing things started to happen . . . our process is producing results faster, better, and cheaper than any other process we have ever witnessed—all in the spirit of Edison's original Invention Factory." He adds, "The secret is learning to see through their eyes."

ELEMENT 23: APPLY THE RIGHT BUSINESS MODEL

During his lifetime, Edison was involved in more than 150 businesses either as an inventor, investor, licensee, licensor, senior manager, or board member. He not only invented products and services, he learned how to run several of his businesses from the ground up. His innovations were targeted to a wide variety of users including businesses, governments, schools, homeowners, and children. The tremendous breadth of Edison's business empire is attributable not only to his skill in designing new technologies for audiences, but also his willingness to experiment with different business models for creating super-value.

The term "business model" refers to the way an enterprise organizes itself to offer products or services to the marketplace. The best business models are designed to create exceptional value for customers. As in Edison's time, contemporary organizations face challenges in establishing appropriate models, and in remaining flexible in response to changing market conditions. Business models are often hard to change because an organization's "way of doing business" becomes ingrained. Commitment, *kaleidoscopic thinking*, and discipline are required to alter or replace a model that may have been successful in the past.

A business model consists of multiple interconnected parts, sometimes called "legs." The name of each leg typically corresponds to an organizational function, like R&D, marketing, distribution, customer service, manufacturing, technical support, data processing, or sales.

A good business model has legs that support the strengths of the organization. All the legs must work together in ways that differentiate the organization from its competitors. If several legs of the business model are not aligned with the organization's strengths, they can drag down overall efficiency just as a high percentage of delinquent loans drags down the overall profitability of a bank.

Thomas Edison began with a business model that emphasized the strengths of his R&D leg. The notion of *having* an R&D leg was itself an innovation. Edison ultimately learned, however, that he needed to have other strengths that complemented his R&D prowess to differentiate his company from its competitors. At first, Edison paired the R&D leg of his business model with strong marketing and manufacturing; and later, following his development of infrastructure to deliver electrical power, with outstanding distribution. He learned that he could profitably use different business models—each emphasizing a different combination of legs—to better serve multiple audiences. Experience taught Edison that it was *how* he combined the legs of his business models that made the difference.

Lamp Works of the Edison General Electric Company at Harrison, N. J.

This illustration of the Edison Lamp Works factory, part of the Edison General Electric Company, shows the massive scale of Edison's manufacturing operations.

To provide super-value for customers, the legs of your business model must work together to deliver benefits that your end-users want, and that reflect your organization's core competencies. In building a base of businesses that could deliver more than twenty different phonograph models and hundreds of record selections to audiences around the world, for example, Edison combined his R&D prowess with marketing, custom manufacturing, and home delivery to create a unique offering that was for music in his day what NetFlix is for movie aficionados today.

Edison's model for bringing lighting to cities and towns across America also began with his R&D prowess, which he used to create the distribution infrastructure required to send light and power out to homes and businesses. In this unique model, he combined his R&D leg with a "construction department" leg—which was responsible for designing and building power plants—as well as a customer service leg designed to serve businesses and municipalities. By realigning the legs of his business model in this way, he created an innovative offering that resembled a franchise operation, bringing forth a model one might call, "Let there be light."

Edison used at least six different business models over the course of his sixty-two-year career. He primarily used the core model shown in the chart on page 194, but he also used five other model variations that will be described shortly. His willingness to experiment with different models and to adjust his primary model based on changing market conditions was a key driver of his innovation success.

The Edison Phonograph was delivered door-to-door by horse-drawn wagon in cities across the nation.

The chart shown on page 194 describes the core business model Edison employed. In the left column are the legs, and in the right column are the activities associated with each leg. It resembles many business models that came before him and after him with the exception of the R&D leg, which Edison introduced. The R&D leg enabled him to drive ideas that proved distinctive in serving his diverse target audiences.

EDISON'S CORE BUSINESS MODEL
UNITED STATES AND INTERNATIONAL

"Legs"	*Activity*
R&D	Develop proprietary products within the Edison laboratory system and protect them through patents.
Commercialization	Launch patented products domestically and internationally, retaining exclusive patent rights.
Manufacturing	Develop custom manufacturing processes for Edison-patented items, conducting manufacturing in owned facilities. Supply product for international orders; establish some manufacturing hubs internationally in select locations.
Marketing	Hire experts to design marketing materials for both business and consumer audiences. Manage the Edison "brand," and train "instructors" on how best to teach customers to properly use Edison products.
Sales	Use trained outside agents who have been "instructed" on how to use Edison products. Pay them salary plus commission.
Distribution	Factory-to-the-public sales with door-to-door home delivery by agents. Catalog sales fulfilled by mail or home delivery. Leased retail space for some products.
Customer Service	Provide technical service and repairs in-house at factory locations. Respond to customer letters and phone inquiries.

Although this core business model is common today, Edison was the first to demonstrate that it could be applied to multiple industries. His success in using this core model for diverse innovations such as the phonograph, the motion picture camera, and the movies—among others—offers evidence of its value and versatility. Edison's ability to leverage his R&D leg enabled him to create a steady stream of proprietary technologies that were practical and profitable.

Here is a graphical representation of Edison's core business model. Using the approach shown below is an excellent way to analyze, or make adjustments to, your own business model.

EDISON'S CORE BUSINESS MODEL—
UNITED STATES AND INTERNATIONAL

Core competencies: Developing leading-edge technology, convergent technology

Legs emphasized: R&D, manufacturing, marketing

Super-value: "WOW from the Wizard!"

R&D	Develop proprietary technologies designed for use in America and internationally.
Manufacturing	Design custom manufacturing processes for highest quality, in owned factories.
Marketing	Hire core staff for brand development and collateral design; train agents and instructors.
Commercialization	File patents and retain exclusive rights. Launch products in the United States and abroad.
Sales	Train agents and instructors to sell Edison-branded goods, services.
Distribution	Factory-to-the-public with home delivery and mail order; leased retail space.
Customer Service	In-house technical support; response to consumer letters.

Here are some of the variations to Edison's core model:

Model Variation #1: Sell or License Intellectual Property. Edison sometimes modified his Commercialization leg to license a technology patented in his laboratory to an outside buyer. This allowed him to generate revenue from inventions that he did not wish to sell or market himself. Often these licensing arrangements also entitled the buyer to use the Edison brand name—a privilege that carried an additional charge. Edison's sale of his electric pen patents to the A. B. Dick Company for use in marketing and distributing the Edison Mimeograph Machine offer an example of this business model variation. He also sold patents related to his dictation machines.

Model Variation #2: License Intellectual Property from Outside Inventors. Edison's core business model reflected his preference to create and market his own technologies. However, as his business empire expanded, he wasn't always able to modify his product lines rapidly enough in response to market changes. In 1895, domestic and international competition became particularly keen in the movie projection industry. Edison's own improvements to the kinetoscope—the world's first movie projector—were lagging. In early 1896, Edison purchased rights to a patented projection system designed by a competitor. He agreed to manufacture and sell the competitor's item under the Edison brand name, while also supplying movies for it until his own improvements were ready. In executing this strategy, he modified the commercialization, manufacturing, distribution, and sales legs of his core model to create an External Licensing function.

Model Variation #3: Develop Franchised Services to Speed Distribution. Edison preferred owning his factories so that he could ensure compliance with his high manufacturing standards. But the tremendous success of his electric lighting distribution system made this strategy too expensive. Instead, he opted for what we would recognize today as a form of franchising. The Edison Construction Department designed power plants for municipalities and other customers, canvassed the desired marketplace, recommended locations for the power facility, then assisted in the building of the plant itself. Often Edison was a stockholder in these ventures, but not always. In this business model variation, Edison adjusted his manufacturing, sales, and distribution legs to create a Franchise Services leg that differed significantly from his primary product-based model.

Model Variation #4: Commercialize Edison Patents Internationally. Edison tailored his inventions to take advantage of international markets. The most significant examples of Edison's use of this model focus on his telegraph and telephone inventions. During his Menlo Park years, while providing technologies to Western Union under contract, Edison developed the carbon button transmitter for use in America, and the "chalk drum receiver" for the United Kingdom. The advantage of the chalk drum receiver lay in its avoidance of Bell's U.K. patents. The combination of Edison's chalk drum receiver and carbon button transmitter system was so successful in England that these patents were later purchased by Bell. Edison benefited by receiving revenue from the cross-licensing agreements that resulted from the purchase and from the strengthening of his brand internationally.

Model Variation #5: Provide Training Services to Build New Industries.   Although Edison's marketing staff instructed its agents on the use of the Edison phonograph, motion picture projector, and other products, the success of Edison's electric lighting distribution system required a huge influx of trained installation, maintenance, and technical personnel. No training resources existed anywhere for this purpose. Edison himself wrote dozens of instruction manuals along with assistant Charles L. Clarke, and in many instances delivered training himself. Although this effort became a significant drain on Edison's time at a point when he was already preoccupied with details of the lighting system, Edison's pioneering efforts to establish what we know today as the field of electrical engineering grew from these seeds. Edison modified the sales, marketing, distribution, and customer service legs of his model to create a Training function.

These five business model variations reveal Edison's willingness to adjust his approaches based on the changing market conditions as well as changing customer needs. Although Edison's approaches were not always perfect on the first try, his willingness to make investments in new ways of doing business allowed him to consistently provide super-value to his customers.

In their book *Ten Rules for Strategic Innovators*, authors Govindarajan and Trimble emphasize that having the right business model is more critical now than ever. They point out that the speed of communication and transportation technologies has heightened the importance of selecting the right business model because the impact of innovation initiatives becomes visible almost immediately. Not only are your senior managers or board of directors watching, your target audience—and your competitors—can review the effectiveness of your efforts via the Internet, cell phone, television, or radio in a matter of hours, or even minutes.

Govindarajan and Trimble stress that when existing organizations desire to create a new business model to drive a major innovation initiative, it's important to borrow the right qualities from the mother organization. According to Govindarajan and

Trimble's research, Edison generally transferred the right kinds of know-how from one model to another, as shown below.

WHAT TO BORROW FROM A PARENT COMPANY WHEN DEVELOPING A NEW BUSINESS MODEL

What Edison Encouraged His Teams to Borrow	What Govindarajan and Trimble Advise Executives Today
—Laboratory experimentation techniques, structure —Edison brand name —Knowledge of manufacturing, marketing	Linkages that lend competitive advantage from the mother operation to the new one
—Persistence —Collaborative team structures —Culture of learning —Open communication and exchange	Core values of the mother organization
—New facilities located within reach by phone, a 4-hour trip by horse-drawn carriage, or under 4 hours by electric automobile	Geographic proximity
—Leaders allowed to hire their own staff —Leaders encouraged to create their own networks	Encourage distinctiveness of the new organization's "DNA"

Edison was able to transfer the strengths of his organization's unique culture across his various business models. When sending members of his inner circle to build a new factory to make parts for his electrical power system, for example, Edison realized these individuals had never undertaken that kind of initiative before. Nor had they ever developed products designed to create infrastructure before. But he knew they took with them the values of his laboratory, the Edison brand name, and knowledge of the core competencies that generated "practical convergent technologies" for the marketplace.

Although Edison maintained close contact with each new operation through its lead manager, he allowed all the new operations to do their own hiring. He got involved in key decisions, but, as in the laboratory, he "turned his boys loose" when giving them assignments. This allowed the most appropriate elements of the existing mother culture to be transferred to the new one.

Established organizations that want to innovate by launching new business models often fail, as Govindarajan and Trimble note, because they borrow too many things from the mother enterprise. If the software systems, hiring practices, reward structures, and planning systems are all borrowed from the mother enterprise, it's difficult for a new culture to emerge. Govindarajan and Trimble emphasize that the appropriate core competencies need to be transferred to the new model, but much also needs to be "forgotten." This is particularly true for planning and forecasting cycles. New business models require more frequent planning and adjustment sequences, allowing an enterprise to plan and forecast even on a weekly basis—particularly if the market is fast-moving. Edison recognized this, and rarely got locked in to specific patterns of forecasting and planning. In the process of building a distribution network for electrical power, many members of Edison's staff gained experience with diverse planning and forecasting techniques, which aided them in later efforts.

Edison also carefully considered how much it would cost to operate each business model. He knew every model had to make financial sense. He quantified his business models by looking at each leg and calculating what it would cost to develop, grow, and maintain it over time. In this process, he recognized that some investments were bigger, and were thus going to take longer to reap rewards—such as his foray into electrical power and his development of the storage battery. In other instances—such as a product improvement for the phonograph—he knew the rewards would come sooner because consumers were already familiar with the basic technology of this invention. He adjusted the structure of his model and his financial projections accordingly.

Edison's approach to crafting the right business model always included an overall assessment of the total market opportunity, as well. Once he had initial figures in mind, he continually refined his quantitative assessments over time. As he concluded a preliminary round of lighting industry cost projections, Edison wrote to a friend in late January 1880: "I have been figuring during the past week on some estimates and they all show that we are going to make enormous profits at the present prices of gas . . ."

Edison and Francis Upton spent almost two years of labor forecasting specific costs for the Pearl Street installation. During the same time period, Edison also forecast the trajectory of the entire industry itself. Ultimately, Edison's thirty-year projection figure came very close to the actual value of the entire electrical lighting infrastructure in 1910.

Carlson and Wilmot sum up this critical element of Edison's success: "To develop your business model you have to do your homework."

Creating Innovation Literacy: Apply the Right Business Model

As Edison understood, creativity is necessary but not sufficient for innovation. Creative individuals often get frustrated when they feel that "management doesn't listen" to their ideas or innovation plans. In many instances, this is because the idea presented does not include all the other ingredients needed for success. If you are *Innovating Like Edison*, **you must craft appropriate business models to support your ideas.** You must consider how to leverage your core competencies and which legs to emphasize. As Edison did, you must consider the market size of the opportunity, and the cost to develop and launch the initiative.

Having an idea is not enough. As Carson and Wilmot stress, "Until . . . there is a viable business solution, there is no opportunity."

Every business model needs to be designed to:

- Deliver super-value to customers

- Align with the organization's core competency(ies)

- Allow innovation to flow through one or more legs of the model in a way that supports differentiation versus competitors

- Capitalize on the total business opportunity

Innovating Like Edison does not mean innovating across every aspect of your business model simultaneously. Edison chose one primary leg—R&D—through which he funneled most of his innovations, supplementing major technology or product innovations with other less intensive innovations in other areas, such as home delivery in his customer service leg. Focus on funneling your innovations through the legs of your business model in ways that maximize your core competencies and leverage the strengths of your organization, allowing you to deliver super-value to your customers as effectively as possible.

Sometimes, the business model itself is the most important innovation. Contemporary examples of innovative business models include:

- **Dell** is a computer company that does not differentiate itself on R&D. It decided instead to emphasize reducing production and distribution costs, and focus on its customer service leg to delight end-users with custom-designed computers. Although Dell has gone through ups and downs, it provides super-value to customers by focusing its business model on distribution and customer service.

- **Southwest Airlines** shunned the industry's traditional "hub and spoke" system, instead focusing on the distribution leg of its business model by targeting popular routes at reduced prices. It combines this with a theme of "air travel can be fun"—a customer service–driven innovation that the company integrates across all its marketing communications. Its primary function, however, is to get passengers on and off Southwest planes rapidly, and to keep its fleet in the air as much as possible.

- **Enterprise Rent-A-Car** disrupted the traditional customer service model of industry leaders Hertz and Avis by offering to pick up customers at their homes—creating super-value for busy travelers who don't have time to get to the nearest rental car location.

- **eBay** transformed the live auction industry's business model by emphasizing its distribution leg, using the Internet to virtually display every item available for sale. eBay then delivered a process innovation featuring proprietary software that allowed customers to bid on items, purchase them, and fulfill orders using online shipment options. eBay's business model bundles groups of online services that create super-value for customers. Its unique approach has paved the way for a new generation of business models coupling online services with a diverse variety of offerings.

Whether you are working in a big corporation or attempting to promote your own inventions, learning to apply the right business model is an essential element in developing your innovation literacy. Dr. Robert Langer offers an inspiring example of how inventive genius translates into super-value through applying the right business model. In 1998, Langer received the Lemelson-MIT Prize for being one of history's most prolific inventors in medicine. In 2003, he won the John Fritz Medal for achievement in engineering, a medal that Thomas Edison won in 1908. Langer was inducted into the National Inventors Hall of Fame in 2006. More than 180 of Langer's patents have been licensed to companies around the world, and many new companies have been launched based on his innovations.

Langer, and his team of collaborators at MIT, have learned the importance of applying the right business model. Despite evidence that most university-based intellectual property development groups are unsuccessful, MIT's Technology Licensing Office has prospered. As Langer describes it,

> MIT's licensing office—called the Technology Licensing Office (TLO)—helps us file patents when we come up with inventions. But almost more importantly, what they also do is create a bridge. The bridge allows us choices—we can either license the patent to another company, or we can form our own company. The TLO does a lot of work to enable a deal to be created that helps the inventors, helps the company, and helps MIT. They make deal structures that are win-win. It's not always necessarily aimed at making money; it's aimed at getting the inventions out there.
>
> In the case of a wafer we developed to treat brain cancer, we actually had sold licensing rights to a small, start-up company, and they actually started doing pretty well with it. It actually got moved to various clinical trials. But then, the small firm got bought out by another company, and the new company basically stopped doing any work on it at all. So, what the TLO said to them was, "We'll give you two choices: Either we're going to take the patent back, or if you put money into it, we'll lower our royalty fees by a factor of ten." TLO will do things like that to make things happen. And the company did actually decide to start yet another company. They put additional money into it—about $2.5 million. Ultimately, the Gliadel® wafer that treats brain cancer came out of that.
>
> I think the success of this kind of business model is hard to put into a formula. But it ultimately comes down to the people, the intellectual property they develop, and a viable bridge into the marketplace. Every deal that the

TLO makes with a company is different. The portion MIT gets is always the same, and I think what we get is 28 percent of what MIT gets. But then, we also have opportunities to act as consultants for those companies, and that arrangement is created between us and the companies. So we have opportunity to generate revenue both from the royalties and from consulting projects.

ELEMENT 24: UNDERSTAND SCALE-UP EFFECTS

Preparing a meal for a few friends is an easy project for most home chefs. Putting on a dinner party for twenty guests is much more challenging. And, cooking for two thousand is a nightmare for anyone other than an experienced caterer. As we've emphasized, a critical element of Edison's extraordinary success as an innovator lay in his willingness to cater to his consumers needs. His passionate concern for the value perception his customers had of his finished products and services led Edison to devote tremendous attention and consideration to understanding scale-up effects. His ability to successfully launch strategic, design, product, service, and technology innovations was dependent on accurately projecting development costs and getting the kinks out of projects *before* they reached the marketplace.

Edison focused on debugging his inventions as much as possible prior to launch. He pioneered the idea of using scale models, prototypes, and pilot launches—or what we now call "soft launches"—to move his products successfully from the laboratory environment into the marketplace. Edison realized that *it is easier to solve a problem when it's small.*

Although scale-up strategies are now considered standard practice in many industries—particularly for new software, computer systems, manufacturing processes, or expanded telecommunications networks—some companies still foist their untested products and services on the public prematurely, often with disastrous results.

By studying three different examples of Edison's approach to scale-up, we can refine and energize our own understanding of this element of super-value creation. These examples range from relatively

This 1912 photograph appeared in an upscale ad for Edison's home movie projector, the first to use safety film that would not catch fire.

easy and inexpensive trial-and-error methods, to a more complicated soft launch, to the construction of complex and costly infrastructure that cannot be tested in any other way than "live" with all equipment in place. Edison and his inner circle used their scale-up experiences to adjust their products to nonlaboratory environments. They also discovered there were almost always unforeseen costs and public relations risks to be considered.

From his earliest days as an inventor, Edison understood the importance of building and testing scale models and prototypes. As he commented to one of his assistants, "A good many inventors try to develop things life-size, and thus spend all their money, instead of first experimenting more freely on a small scale." Edison had three-dimensional models of every invention built to scale. Rigorous tests of his prototypes offered Edison the opportunity to spot defects and make improvements, saving significant amounts of money and effort, as well as increasing value for the consumer.

One of the hallmarks of Edison's innovation approach was his commitment to reduce the requirement for skilled labor to operate his equipment. Whether it was a telegraph machine or a phonograph record, he constantly sought ways to make his products easy to use and operate. Edison's "ease of use" philosophy aided him in planning and executing scale-ups from the beginning of his career.

As Edison had very little money to invest in the process of developing and testing his inventions during his years as a telegrapher, he tapped his coworkers to help him with the debugging process. With help from friends in the machine shop, Edison built simple prototypes of his new, improved telegraph equipment. Edison knew that existing telegraph lines would be fine for his purposes, and he had a ready supply of fellow telegraphers who were eager to help him test the equipment. He carefully logged the results of his trials, and applied what he learned to perfect his innovation. Edison's diligence in the scale-up process allowed him to approach clients with full confidence in the efficacy and reliability of his new telegraph equipment.

The success of his new telegraph equipment made it possible for Edison to become an independent inventor and this set the stage for the success of his system for providing incandescent electric light. The lighting system was the result of dozens of inventions based on thousands of experiments, and then, a series of increasingly complex scale-up efforts. Prior to illuminating a one-mile square area of New York City, Edison began by lighting his home and laboratory at Menlo Park, New Jersey.

When his breakthrough with incandescence was trumpeted by the press in December 1879, Edison was swamped with offers for testing his lighting system on a broader scale. He rejected the proposals, despite the potential public relations benefits, because they didn't offer sufficient opportunities to evaluate the system's efficacy as thoroughly as he believed necessary. Then, in early 1880, mogul Henry Villard—a friend of Grosvenor Lowrey's—informed Edison he was building a new steamship called the *Columbia*, and would be docking it in the New York harbor for outfitting. Villard had attended Edison's first demonstrations of house-to-house Christmas lighting in December 1879 at Menlo Park, and had become an immediate enthusiast. He proposed that Edison test the lighting system on his ship. Edison accepted Villard's offer because he believed at least half of the systems required for the *Columbia* soft launch would parallel those needed for a major urban system; he also felt that a victory with the *Columbia* installation would be a public relations bonanza that would help deter competitors.

As he prepared for the soft launch of the *Columbia* lighting system, Edison pressed Upton, Batchelor, Insull, and others to refine the dynamos, insulation, wiring, switching, and other systems required for a larger-scale urban launch. The scale-up experiment on the *Columbia* helped accelerate development of several key features of the full-scale system, particularly the first lamp sockets with key-switches, as well as "safety wires," i.e., fuses, that were the main protection against fires caused by short-circuiting.

When the *Columbia* was ready to sail in late 1880, Francis Upton took charge of its new lighting system, and was aboard ship making detailed records of every aspect of its functioning. The *Columbia* was the first vessel to sail with a fully working onboard system of electrical illumination.

Although feedback from the *Columbia* soft launch was positive, it generated some unintended media spin. A few reports suggested that incandescence was suited only for specialized, limited markets. This spurred Edison and his inner circle to redouble their efforts to find an appropriate venue for a larger-scale public launch.

His trial calculations suggested that his new central power system would be best suited for a high-consumption service area of roughly one square mile. After months of searching for a perfect location, Edison found it in Lower Manhattan near Wall Street, in an area now called the Pearl Street neighborhood. In the fall of 1880, New York's Board of Aldermen agreed to allow Edison to test his central power system in this neighborhood. Edison knew the influential New York media would be watching his every move. He also understood the benefits that might accrue from successfully illuminating an area so close to Wall Street, New York's financial center. Edison set an aggressive launch date of September 4, 1882.

The Pearl Street test of Edison's new system required construction of completely new infrastructure, and the adaptation of hundreds of patented inventions rigorously tested at Menlo Park and on the *Columbia*. In addition to perfecting the filament in the incandescent bulb itself, Edison and his team worked feverishly to complete the myriad of streetlights, meters, switches, underground conduits, and complex circuitry required. However, some equipment—like the thirty-ton dynamo—was not practical to construct, test, and then move. The full test of the complete system had to take place "on location."

In many ways, installing the first electrical power system in 1882 was like running the first OS/2 computer operating system in the 1990s: trying to debug an entirely new-to-the-world system without sufficient numbers of trained software engineers. Edison found his installation work "seriously retarded by an inability to obtain competent engineers." His solution was to set up a training school for installers in New York City.

Despite the extraordinary infrastructure and training challenges Edison faced, on September 4, 1882, all was ready. Edison synchronized his watch with station engineer John Lieb, and "accompanied by Kruesi, Bergmann, and others, made his way to the offices of J. Pierpont Morgan in the Drexel building at Broad and Wall streets . . . There he supervised the installation of the safety catches and at three o'clock in the afternoon turned on the office lamps." The "switch was flipped," and incandescent light bulbs across the one-square-mile region of the Pearl Street neighborhood flickered on. The next day, the *New York Herald* reported the historic event:

> In stores and business places throughout the lower quarters of the city there was a strange glow last night. The dim flicker of gas, often subdued and debilitated by grim and uncleanly globes, was supplanted by a steady glare, bright and mellow, which illuminated interiors and shone through windows fixed and unwavering.

The glowing reviews by the press helped soften the bite of the bill Edison and his investors paid for the Pearl Street scale-up: a price tag approximating $171,400 ($3.2 million today). Despite the financial challenges, Edison felt it was worth two full years of intense labor to achieve success. His biggest surprise and most important lesson concerned the challenge of training enough skilled workers. He vowed to create enough trained workers so that the industry itself could thrive. Ultimately, Edison helped fund a new program at Columbia University by donating $50,000 ($929,000 today) of equipment from the 1881 Paris International Electrical Exhibition, making Columbia University one of the first American institutions to offer a degree in electrical engineering along with MIT and Cornell.

The Pearl Street scale-up also yielded another key realization for Edison: A different type of system would be required to serve less densely populated areas. The central power station configuration designed for Pearl Street worked only for a densely populated community. Based on his learning from Pearl Street, Edison immediately began designing a second power system configuration called the "three-wire system," for smaller communities.

The scale-up Edison completed for this world-changing strategic innovation—one of the most complex of all time—demonstrates the vision and tenacity required to build and scale-up new-to-the-world systems, particularly those involving new infrastructure. We can take lessons from Edison's experiences by ensuring that scale-up becomes a regular part of the innovation process, and that allocations of money and time allow for the unexpected results that can emerge.

Creating Innovation Literacy: Understand Scale-up Effects

Understanding scale-up effects is an important and often overlooked aspect of innovation literacy. If you want to deliver super-value, you must attend to this pivotal element.

Kazuhiko "Kay" Nishi joined Microsoft in 1979, and became vice president in charge of new technologies before leaving in 1986 to become the leading figure in Japan's personal computer industry. He comments on the importance of understanding scale-up effects: "There are two types of creativity: the creativity of making zero to one, and the creativity of making one to a thousand." Carlson and Wilmot add that "moving from one to a thousand requires a long journey and most of the hard work is invisible in the beginning."

Dr. Helen Free, an inductee to the National Inventors Hall of Fame in 2000, describes how she and her husband took their creativity from "one to a thousand" and beyond. In the 1950s and 1960s, they were instrumental in establishing the market for urinalysis and blood droplet diabetes testing, pioneering the in-home self-testing industry and paving the way for in-home pregnancy testing. Helen was involved from initial concept all the way through scale-up and launch. She describes the process:

> In the 1950s, my husband and lab director, Dr. Alfred Free, devised the wonderful "double sequential enzymatic reaction" that provided the foundation for in-home testing. He patented it, and this basic patent was a breakthrough because the process itself could be done with an easy "dip and read" test.
>
> From the beginning, we realized we could impregnate into paper a reagent that measured glucose. This was back in the days when we did everything in the lab from scratch—we had to weigh out the chemicals on a balance, and mix them in volumetric flasks. It was all manual. We made big sheets of treated paper, hung them up to dry, and we'd cut individual squares out . . . Then, Al had an Edison reaction I guess, and he said, "You know, if we can do it on squares of paper, why don't we just make it in strip form?"
>
> Although we had manufacturing experts within the Ames Division of Miles Laboratories (now Bayer) who assisted with the scale-up and commercialization, we lab people devised the actual manufacturing methods to make the product. From the beginning to the commercialization, we were involved. To prepare for the manufacturing step, we hung up these big sheets of paper in an oven. And when they dried, we cut them with scissors and put them in bottles . . . testing their efficacy at increasing levels of production. The first product was Clinistix® which we made right here in Elkhart (Indiana) in 1956.

Dr. Free's success story illustrates a very important lesson: **Scale-up works best when a multidisciplinary team collaborates on the process.** In other words, people who don't normally work together need to connect as early in the innovation process as possible to make the scale-up run smoothly. Dr. Free's in-house manufacturing team observed as she and her colleagues successfully produced their "homemade" batches. This allowed the manufacturing experts to isolate key elements that would drive the greatest efficiency in scale-up. Together they devised a mass production process that could churn out tens of thousands of Clinistix batches flawlessly. And, they included Marketing early in the process as well, linking the scientific achievement and production process with a powerful, appropriate strategy for communicating the benefits of this revolutionary technology.

Tom Quick is vice president for quality at Spectrum Brands, manufacturer of Remington electric razors and a host of other products. Tom has scaled-up hundreds of brands, and has attempted to systematically apply the same principles that Dr. Free and her team developed decades ago. He coordinates a variety of in-house resources to ensure scale-up proceeds in as time-efficient and cost-effective a manner as possible. Tom describes his approach:

> Spectrum Brands uses what's called a "stage-gate" new product development process to move projects from concept all the way to production. "Stage-gate" means that new products are monitored in specific ways to meet the thresholds we've established for success. There are two key methodologies we use to scale-up from prototype, to engineering builds, to plant trials, to mass production. These are called "design verification" and "production validation." Very simply, verification is: "Can we make it once?" Validation is: "Can we make it a million times?" A profit-and-loss projection for the whole effort must be approved early in the development of the new product including projected costs for scale-up. There are specific target dates set out for scale-up in the beginning of the process, tied to launch.
>
> The first step after successful "Engineering builds" (one or two samples) is a plant trial. The plant trial is usually a couple hundred units meant to test the production systems. There is usually a second plant trial at about double the volume. The next step is limited production; it is the first saleable (acceptable) product. It may not be quite perfect, and therefore the production is limited in some way. For the final release of the product, Manufacturing is authorized to make what quantities are necessary. At each step there are prescribed tests that must be completed. Costs, both unit costs and investment (tooling), are tracked very carefully.
>
> Verification is "owned" by Engineering, and they must show that the design meets the agreed specifications (specs). Validation is "owned" by Quality, and they must show that what is made on the production line meets the final specs that came out of verification. The whole process is very efficient in that Engineering is "pushing" while Quality Assurance is "pulling," with both groups working hard to get the best product out the door. In this manner, we successfully launch hundreds of new products every year.

Quick adds,

Scale-up is about adding more and more sources of variability till you end up exposed to all possible sources. Success means facing all this variability and still meeting the requirements of our customers. This is the only way to achieve maximum value. You must start and end with the customers' requirements, and focus on them in each step of the scale-up process.

The principles that Tom Quick discusses are just as relevant in a service business. Before launching a new service, walk through all the potential scenarios your new or existing customers will face when you "go live." When a new customer wants to buy from you, or simply seeks to find out more about your services, they're experiencing the results of your "scale-up" efforts. Whatever industry you're in, scale-up is about solving problems when they're small so you can deliver big value to your customers.

Of course, infrastructure scale-ups, like Edison's lighting distribution system, are typically much more time-consuming and expensive than most product or service scale-ups. This is why governments and other bodies involved in large projects need to be just as aware of scale-up processes as teams who do it routinely every year. With its roots in DARPA—the U.S. Defense Advanced Research Projects Agency—what we know today as the Internet required roughly twenty years to scale-up. Dr. Robert Kahn, an MIT faculty member who joined DARPA in 1972 and remained active there for thirteen years, collaborated with colleague Vinton Cerf, a UCLA Ph.D. who was then an assistant professor at Stanford, to create the architecture for the Internet, pioneering the design for what we know today as the TCP/IP Protocol. The TCP/IP Protocol now serves as the standard host protocol enabling applications ranging from e-mail to instant messaging on the World Wide Web. In a recent interview with the authors, Dr. Kahn commented:

If you go back to the 1972–73 time frame when DARPA first started thinking about connecting multiple nets, it was arguably twenty years before what we now know as the Internet was fully scaled-up. Infrastructure takes a long time to deploy and take hold. It's much different from most consumer products or services in this respect.

Robert Kahn and Vinton Cerf were inducted into the National Inventors Hall of Fame in 2006 for their landmark achievement in developing the TCP/IP Protocol.

ELEMENT 25: CREATE AN UNFORGETTABLE MARKET-MOVING BRAND

> All the world's a stage . . .
> —*William Shakespeare*, As You Like It

Edison's skill at generating insights and linking his core competencies with market gaps was complemented by a finely honed focus on tuning in to his target audience. He cultivated his ability to apply the right business model for his various innovations, and he continually deepened his understanding of the most effective approach to scale-up. All these elements made it possible for him to translate his inventions into innovations, and deliver high-quality products and services at exceptional value. But, Edison also understood another essential, though less tangible, aspect of *super-value creation*: managing the perception of his efforts.

In the late 1800s and early 1900s there were few marketing texts or branding consultants to guide Edison's efforts. Yet, Edison persevered in his intention to create an unforgettable market-moving brand using many tools we would recognize today, including a heavy emphasis on public relations, live product demonstrations, and targeted collateral materials. He captured significant "share of mind" with diverse audiences, including what we now call business-to-consumer as well as business-to-business customers. Operating without the benefit of computers, the Internet—or even radio and television until the turn of the twentieth century—Edison created an international marketing colossus enviable even by today's standards.

For both consumers and business owners, Edison's name became synonymous with leading-edge technology, novel product applications, superior quality, variety, and value. But beyond these quantifiable aspects, Edison always cultivated a certain "je ne sais quoi." He crafted a unique, magical aura for the Edison brand by melding his image as a "man of the people" with the perception of him as a genius who embodied progress through science and technology.

Consumers and businesses were confident that products bearing the Edison name would be of the highest quality and durability, and investors felt that the ". . . Edison name had come to serve as a form of assurance for untried new technologies." The market anticipated the breakthrough

Edison's image often appeared on labels for his products, assuring customers they were receiving "the genuine article." This label for one of Edison's movies lists the years of his film patents, as well as locations for film manufacturing operations and dealerships in New York, Chicago, San Francisco, London, Paris, and Berlin.

nature of Edison's products, expecting a high threshold of performance with each launch.

A marvelous example of the power of the Edison brand in the market is provided by film concessionaires Norman C. Raff and Frank R. Gammon in their response to one of Edison's competitors, just prior to the release of Edison's motion picture camera:

> No matter how good a machine should be invented by another and no matter how satisfactory or superior the results of such a machine invented by another might be, yet we find the great majority of the parties who are interested and who desire to invest in such a machine have been waiting for the Edison machine and would never be satisfied with anything else, but will hold off until they found what Edison could accomplish . . . evidently believing that Edison would in due time perfect and put out a machine which would cast the others in the shade.

Edison worked throughout his career to enhance the magnetism of his brand. His sensitivity to consumer needs and perceptions was facilitated by his experiences as a teenage newspaper publisher. As biographer Lawrence Frost comments, "Though the fifteen-year-old publisher had his problems with grammar and spelling, he knew what his readers' interests were."

His newspaper experiences helped Edison understand what reporters and publishers were seeking, and he learned how to deliver it. The press loved him. As one reporter remarked, "Edison is the Aladdin's lamp of the newspaper man. The fellow who approaches him has only to think out what he wants to get before taking the lamp in his hand and he gets it."

The media loved to publish photographs of Edison laboring outside wearing dungarees and work clothes, or in a chemical-splattered suit inside the laboratory. They knew Edison's image resonated both with the laboring man who saw him chewing tobacco or smoking a cigar, as well as the intellectual reader who valued his extraordinary insights.

Edison's "common touch," humor, penetrating intellect, and broad range of interests meant that not only did the press enjoy interviewing him when a topic was "hot," but they could turn to him whenever they needed a story. Reporters knew Edison was always ready to help them with a "scoop" on his forthcoming products, or his predictions on what lay ahead for the U.S. economy. He could summon reporters to his West Orange laboratory and have a throng of newsmen at his doorstep within an hour.

Edison's "homespun humor and forceful statements . . . helped make him an attractive subject for reporters . . ." Edison often made brash predictions about what he intended to achieve. As one reporter noted, "He is fond of startling his hearers with extraordinary statements, but so equally extraordinary are the things he has done that one is wise not to venture too far in pronouncing between fact and fancy."

Edison leveraged his contacts at newspapers, magazines, and journals through-out the United States and abroad. When radio was introduced, Edison was quick to take advantage of the new media, delivering broadcasts from his West Orange laboratory, or while vacationing at Chautauqua. Late in his career, he began offer-ing reporters one-on-one access every year on his birthday, February 11. The pub-lic could listen to the "Wizard of Menlo Park" in live conversation with a series of top interviewers throughout the day. Eventually, February 11 was declared Na-tional Inventors Day in Edison's honor.

Then, with the introduction of the Edison Motion Picture Camera, interviews with Edison were sometimes captured on film, showing Edison alongside presi-dents, foreign dignitaries, scientists, and other notable figures. Edison used the new media he had helped to create to catapult the mystique of his brand name to new heights, cementing his image as a living legend. These appearances did more than magnify awareness of the goods and services Edison had created through his laboratories; they seared Edison into the minds of the consuming public as a na-tional figure central to the well-being of the nation's economy.

Edison's skill with the press also paid great dividends in helping him overcome the effects of his mistakes. His worst moment in the media came as a result of his attempts to discredit his rival George Westinghouse. Edison was roundly criti-cized for his campaign against his competitor's AC power, including his support for an AC electric chair. Edison lost the battle to make DC the industry standard because AC power was proven superior. Nevertheless, because of Edison's unprecedented successes and his strong relationship with the media, the press ultimately aided Edison in maintaining a strong, positive reputation with the public.

Edison's brand-building strategy included more than just masterful media rela-tions. He integrated his marketing and branding efforts at every level to manage the way his endeavors were perceived. With the help of his "marketing director,"

The magic of the Edison brand was cultivated with ads like this one for the new Edison Phonograph, positioning it as an item invented, perfected, and manufactured by Edison.

Edward H. Johnson, Edison created an approach that integrated all the communications reaching consumers. For example, the collateral materials Edison used—such as brochures, instruction booklets, and advertisements—often used big, sensational headlines; but these attention-grabbing headlines were always followed with more down-to-earth language focusing on the real needs of the consumer or business audience. Photography and illustrations used in the collateral materials featured images of consumers using the product, portraying them in ways designed to appeal to their aspirations and ideal self-image. Despite the cutting-edge nature of the products themselves, the language was focused primarily on the experience of the user rather than the technology. In all their marketing materials, Edison and Johnson emphasized ease of use, enjoyment, and long-lasting benefits available to buyers. This combination communicated super-value to customers.

Edison's integrated marketing approach also included the use of hired salespeople trained as "instructors." Instructors sold Edison's products through retail outlets, took orders, then arranged for the products to be delivered to the customer's home or place of business via horse-drawn wagon. Product training for consumers was a particularly important differentiator of Edison's brand versus his competitors. Edison insisted that consumers be educated about what they were buying. In the early days of the phonograph, Edison himself helped new customers understand how to gain maximum benefit from his "baby."

Product education, in Edison's view, represented a significant way to enhance his customers' value perceptions of his products. Edison knew that if his instructors did a good job, then the customer would not only enjoy the product but would also have a stronger relationship with the Edison brand.

Edison reveled in the ingenuity of his creations and loved nothing more than to demonstrate how they worked. His delight was transferred to customers through his sales team as well as through creatively staged product demonstrations. Edison used these demonstrations to generate consumer interest in an item even before it was launched. In his campaign to promote his storage battery, for example, Edison summoned his media contacts to watch as automobiles, powered by his new battery, drove up steep hills near the West Orange lab. A consummate showman, Edison highlighted the resilience of his design by tossing a few batteries out of a third-story window. The press loved it, and Edison followed up with a targeted mailing campaign.

Edison used pricing, customer service, quality control, and design strategies as other tools to build his brand. Edison's new-to-the-world technologies were expensive to produce and could command premium prices. But Edison didn't want to be perceived as a purveyor to an exclusive elite, so he often offered multiple varieties of an item at various price points. This was particularly true of the phonograph, which Edison desired to make available to every household, offering more than twenty different models across a wide price range. For businesses, Edison also offered his products in different sizes and price ranges. Edison introduced six different sizes of dynamos and three different storage batteries to bring them within reach of more companies. Competitors lacked the depth of technological skill to offer so many different items, and even if they *could* keep up technologically, they found it very difficult to compete with the magic of the Edison brand name.

Edison also appreciated the role of design in strengthening his brand. During his years as a telegrapher, he watched fellow operators struggle with overly complex equipment. He wanted all his products to be easy to use. And, he knew that the right look and feel would make customers want to show off their purchases in homes and businesses. Edison wanted people to feel a sense of pride in their ownership of all his products and he knew that, in this

WHICH
is the best
phonograph?

Decide this question for yourself. The first scientific comparison of phonographs has been installed in our store. Hear the

EDISON TURN-TABLE COMPARISON

Edison believed customer education was critical to brand growth. He encouraged prospective customers to compare the quality of his products with those of his competitors, as illustrated in this early twentieth-century phonograph advertisement.

"MASTER BRAND" STRATEGY

Edison created what we now call a "Master Brand" to solidify his market leadership in as many categories as possible. A Master Brand is a brand name that acts like an umbrella, covering a diverse array of industries and price points. Although this is not an easy strategy to adopt, it is one that worked well for Edison. Sony, IBM, Virgin, Disney, Hewlett-Packard, Nike, and Apple use a similar strategy today.

regard, aesthetics was as important as functionality. Among the many beautiful and useful designs Edison introduced, perhaps the most notable is the glass lamp for the incandescent light bulb; its ergonomics and aesthetics have withstood the test of time.

Edison's pricing and design strategies were backed up by technical support and quality control initiatives. He knew consumer perceptions of his top-quality brand name were dependent on top-quality technical support. He instituted a product guarantee: If a customer was dissatisfied with an Edison product, it could be returned or substituted for another. Although there were no written "service warranties" in Edison's day, returns or exchanges were based on providing either a purchase receipt, or simply one's word.

Edison stood behind every product bearing his name. His ability to guarantee his products was predicated on the stringent quality controls he applied to every stage of his design, manufacturing, and sales processes. If a flaw managed to escape Edison's scrutiny, he would do—and spend—whatever was necessary to fix it.

Although Edison had a remarkable feeling for his customers' perceptions of his products and their design and functioning, he wasn't as perspicacious in his assessment of their tastes in entertainment. The Edison brand suffered from his stubborn insistence on maintaining artistic control of the musicians and vocalists appearing on his record label. He wanted his records to have a more educational spin, but the public was more interested in pure entertainment.

Edison's ability to gauge the kinds of music people wanted to hear or the types of movies they wanted to watch wasn't on par was his other market-moving brand initiatives. Moreover, the sale of his stock in GE and his inability to keep his name in that company's title—along with his ten-year investment in his less-than-stellar ore-mining business—all served to compromise the continuity of the Edison brand. Nevertheless, his integrated use of public relations, live demonstrations, collateral material, customer training, user-friendly design, product support, and stringent quality control all led to super-value creation.

CAVEAT VENDITOR

Professor James Clawson comments on creating an unforgettable market-moving brand:

All of the great innovators who have become market-moving brands—Disney, Coke, Hewlett-Packard, Nike, etc.—have grown up on the foundation laid by Thomas Edison. While they may not all have studied his life and career, they surely benefited from the processes that he demonstrated to the world during his remarkable lifetime. Edison showed that market-moving brands are developed *first* by their extensive and universal utility, not by the Marketing department. Edison's innate scientific and intellectual curiosity, combined with his determination to measure success based on how customers adopted and used his products, is a rare and powerful combination.

Today, problems arise from the minds of those who see global market-moving brands and want to create new ones, but focus on "sales" instead of "creating value." Those who focus their managerial attention on the sales function and ignore the causal factors that create sales do so at the risk of their company's reputations. Edison was dedicated to building customer satisfaction into every aspect of his innovation process. He went back even further than the causal factors of sales, and saw that "customer value perception" and *its* precursors—dependability, quality, cost, efficiency, and utility—were critical to marketplace success. By focusing on customer value, Edison was able to generate sales as a secondary, downstream by-product, and profits followed.

Those individuals or executive teams who try to build a brand through marketing alone end up on the trash heap of business history.

Creating Innovation Literacy: Create an Unforgettable Market-moving Brand

> The play's the thing.
> —*William Shakespeare*, Hamlet

Thomas Edison grew up with a deep appreciation for the dramatic through his love of Shakespeare. Although he never realized his fantasy of becoming a Shakespearean actor, Edison applied the principles of theater to create his unforgettable market-moving brand. You can apply those same principles to improve your ability to create super-value for your audience. The principles of theater are relevant to every presentation, whether internal or external, and to every advertising and marketing effort. The principles also apply to your efforts to gain support for an innovative idea, and to build your own personal brand.

As legendary business consultant Tom Peters and others have suggested, in a world of hyperchange, you must "be your own brand." In other words, you must develop a strategy to ensure that you are perceived as a purveyor of super-value to whatever organization engages your services. The principles of theater will help you build your own brand just as they can help you build the brand of any organization that employs you.

So, what are these principles? Well, we've already introduced one of the most fundamental principles in our discussion of Element 22 (Tune In to Your Target Audience, page 185). It's so important that it's worth considering it further. Shakespeare understood that for his plays to be popular, they must amuse, move, enchant, inspire, and entertain his audiences. He knew the only way to measure his success as a playwright was to consider the response of his public; just as Edison knew he could measure the success of his inventions only if people wanted to buy them. The simple but essentially important underlying principle is: **You can measure the success of your communication only by the result you get with your audience.** This should be common sense, but instead it's commonly ignored or forgotten by communicators at all levels, from self-referencing sharing in intimate relationships (i.e., someone who repeatedly brings the conversation back to themselves) to molecule-fondling in innovation proposals.

Edison's love of Shakespeare shines through in this charming advertisement for a tabletop version of Edison's phonograph, priced to be affordable to the average American household.

The corollary of the first principle is: **To get results with your audience, they must remember your message.** Again, this seems like common sense, but it is commonly forgotten because many people make the false assumption that "understanding" is the same thing as "remembering." It's not! You can sit through a very well organized presentation, or watch a very expensive commercial on television, and understand everything in it—then forget it completely. The same thing often happens with branding initiatives; they can be expensive, clever, well organized, yet utterly forgettable.

How can you be sure that your audience remembers your message in addition to understanding it? As Edison knew, all the keys to making your message unforgettable are contained in Shakespeare, and contemporary research into the psychology of memory has confirmed these keys to recall. We will guide you to understand and remember the keys for getting your audience to remember, but first let's consolidate this fundamental principle and its corollary in terms of practical application to getting the results you want with your audience.

In planning any type of communication—whether verbal, written, or electronic—begin by setting clear objectives for what you want your audience to *know, feel,* and *do* as a result. In formulating these objectives apply the KISS principle (Keep It Simple, Shakespeare). If you are attempting to gain support for an innovative idea or project, for example, be sure that you know exactly what you want your audience to remember about it. Avoid molecule-fondling and keep your content as brief as possible. Edison was a master at giving his audiences an experience of his products and services that made them feel they were part of something special. He knew that people buy on emotion, and then support their gut response with facts. He captured minds and hearts with his advertising, product demonstrations, collateral materials, and through his own live presentations. And, he always focused on exactly what he wanted his audience to *do* as a result of his communication.

Once you know exactly what you want your audience to remember, you can then apply the principles for making your message memorable. You can discover these principles for yourself by trying this simple exercise:

Below are 50 words. Read through them one time only—left to right. Don't study them, just read each word in turn, or have someone read the words to you. Take no more than ninety seconds to go through the words. When you're done, write down the words you remember.

Snow car pole deck table bottle light sand sky book soap spoon plant rug Edison cellar gate pillow trunk paper road knife stool hay note air rain bird innovation string zone coat cup *Mickey Rooney* light wind tree pencil rope stamp tape light coal card pick truck cape pilot desk frame.

Almost everyone who completes this exercise discovers that they remember the first two or three words from the list (snow, car, pole). Psychologists refer to our tendency to remember what comes first as the "Primacy Effect." Most people also remember the last few words (pilot, desk, frame). Researchers call the phenomenon of remembering what comes at the end of a sequence, the "Recency Effect."

In between "Primacy" and "Recency," however, a lot is forgotten. In the case of our fifty-word memory exercise, the exceptions usually are the word that was repeated three times (light), the outstanding or unusual word (Mickey Rooney), and the words that have a special personal meaning in the context of the exercise (Edison, innovation). Research into memory confirms that, in addition to remembering the first and last elements in a sequence, we also tend to remember anything that is repeated, outstanding, or personally associated.

PROPAR is a acronym to help you remember the five keys to designing communications that your audiences will recall. PROPAR stands for **P**rimacy, **R**epetition, **O**utstanding, **P**ersonal **A**ssociation, **R**ecency. Let's consider how these five keys apply to all your communications and your ability to *Innovate Like Edison:*

Primacy: Edison's favorite play (Shakespeare's *Richard III*) begins with a mesmerizing soliloquy in which the deformed Richard describes his intention to murder his brothers and nephews to ascend to the throne. The audience is "hooked" immediately. In business, the same principle applies. Communicate benefits and "hook" your audience in the beginning. People remember "first impressions," and you don't get a second chance to make one.

Repetition: Richard, who was the Duke of Gloucester with aspirations of being king, repeats the word NOW twice in the first few lines of his soliloquy to make us join him in the present moment. He then repeats the word OUR to begin three straight lines so we feel a sense of connection to events in the kingdom. And then he uses the words "I," "my," or "mine" twelve times in the remainder of the opening, so we must remember that what follows will be all about his own twisted, narcissistic but utterly fascinating desire to become King Richard III. Shakespeare knew that we remember anything that is repeated. So did Edison. He took every opportunity to reiterate the benefits of his products and he did it through every available form of media. Don't assume that your audience will remember your message just because they seem to understand it. Repeat your key points every chance you get.

Outstanding: Edison's favorite play is filled with outstanding, unforgettable scenes like Richard's seduction of Lady Anne, which takes place shortly after he's murdered her husband. Richard exults, "Was ever woman in this humour wooed? Was ever woman in this humour won? I'll have her, but I will not keep her long." Shakespeare makes his main character outstandingly reprehensible so we will not be able to forget him, or the play. Edison also knew the importance of making his message outstanding through drama. When he wanted to convince reporters about the efficacy of his storage battery he didn't just show them the specs and research data; he threw batteries out the window and used them to power cars up steep hills to make his message outstanding. Edison knew that every business presentation is a form of theater, whether his audience was reporters, investors, customers, or his own Muckers. Use your imagination, as Edison did, to generate creative ways to make your message outstanding.

Personal Association: Shakespeare's audience included people from all walks of life and all social classes. How was he able to reach "groundlings" and nobles with the very same play? Shakespeare tuned into the common elements of human experience that cut across all external distinctions. He knew, for example, that everyone sometimes feels wounded and spiteful, and he gives us Richard as an exaggerated mirror for this all-too-human quality. Thus, we can relate to Richard even as we despise him. Like Shakespeare, Edison also reached a remarkably diverse audience. He knew that we remember things that are relevant to our needs and desires. Edison used unpretentious, everyday language to describe his state-of-the-art technology. His advertising and collateral materials featured images of "regular" people using his products and services. Like Edison, tune in to your target audience. Use the terms and methods that allow your audience to make personal associations with your message.

Recency: *Richard III* ends with a beautiful poem, affirming a vision of peace between the houses of York and Lancaster. Shakespeare wanted to leave his audiences with a positive feeling for dessert, after serving them an appetizer and main course of intrigue and treachery. After entertaining his audiences with jokes, stories, and compelling dramatic demonstrations, Edison was confident that they knew what he wanted *them* to know, and felt how he wanted *them* to feel. He never lost sight of the purpose of his communication: to get his audience to do what he wanted *them* to do.

In the case of an individual presentation, you bring it together by focusing on communication that will help you achieve your immediate objective. Perhaps you want:

- Financing

- The opportunity to give another more in-depth presentation

- More staff or support services

- Freedom from normal rules

Whatever you are asking for, be sure to finish by giving your audience the best possible opportunity to give it to you. As you focus on accomplishing your short-term objective, bear in mind that all your communications must serve your larger goal of strengthening your brand.

Your ability to influence others in a memorable way, as Edison did, will dramatically improve your prospects for success in every endeavor. Setting clear audience-centered objectives and applying the PROPAR principles will be invaluable assets in establishing and developing your own personal brand.

And, if you have a creative idea or invention and you want to turn it into an innovation, your communication skills will probably make or break you. As Carlson and Wilmot from SRI emphasize:

> If you are the champion of a new innovation, one of your jobs is to raise the financial and human resources needed to get your project completed. This always means that you must convince someone—a company president, a board of directors, a venture capitalist, or a government program manager— that you have a good idea. In most situations, the number of possible projects far outstrips the available financial resources.

At SRI Carlson and Wilmot encourage innovation champions to develop what they call "Elevator Pitches"—succinct, pithy, PROPAR presentations designed to

THE EDISON INNOVATION LITERACY BLUEPRINT™
5 Competencies, 25 Elements

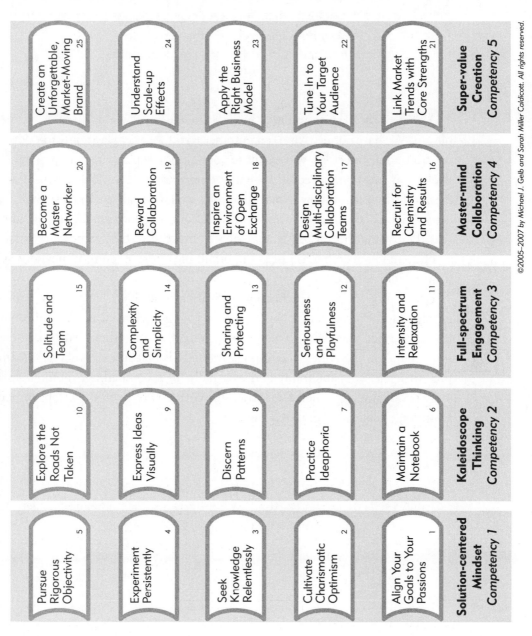

Solution-centered Mindset **Competency 1**	Kaleidoscope Thinking **Competency 2**	Full-spectrum Engagement **Competency 3**	Master-mind Collaboration **Competency 4**	Super-value Creation **Competency 5**
Align Your Goals to Your Passions — 1	Maintain a Notebook — 6	Intensity and Relaxation — 11	Recruit for Chemistry and Results — 16	Link Market Trends with Core Strengths — 21
Cultivate Charismatic Optimism — 2	Practice Ideaphoria — 7	Seriousness and Playfulness — 12	Design Multi-disciplinary Collaboration Teams — 17	Tune In to Your Target Audience — 22
Seek Knowledge Relentlessly — 3	Discern Patterns — 8	Sharing and Protecting — 13	Inspire an Environment of Open Exchange — 18	Apply the Right Business Model — 23
Experiment Persistently — 4	Express Ideas Visually — 9	Complexity and Simplicity — 14	Reward Collaboration — 19	Understand Scale-up Effects — 24
Pursue Rigorous Objectivity — 5	Explore the Roads Not Taken — 10	Solitude and Team — 15	Become a Master Networker — 20	Create an Unforgettable, Market-Moving Brand — 25

To aid you in visualizing Edison's Five Competencies of Innovation and the twenty-five elements that support them, we've created a summary chart called the Edison Innovation Literacy Blueprint. The Five Competencies of Innovation are listed horizontally along the bottom of the chart, and the elements rise vertically above them. Use the Edison Innovation Literacy Blueprint for reference as you begin *Innovating Like Edison.*

elicit support for innovative value propositions. They counsel application of Shakespeare's reminder in *Hamlet* that brevity is the soul of wit.

> All the world's a stage,
> And all the men and women merely players.
> They have their exits and their entrances,
> And one man in his time plays many parts. . . .
> —*Shakespeare*, As You Like It

All the world's a stage, and that includes the business world. All of us are merely players. We have our exits and our entrances, and we may play many roles over the course of one career. As you cultivate your ability to play your many parts effectively, you must also become skilled at contributing to various ensembles. The best performers know how to play their roles well while always making their fellow actors look good. Knowing what you want your audiences to *know*, *feel*, and *do*, and then using the PROPAR approach to make sure they remember what you want them to remember, is just as important in your team communications as it is for your individual efforts.

Everything you've learned about cultivating innovation literacy a la Edison is designed to help you deliver super-value to your audience, as he did. Edison carefully cultivated talents in showmanship brought all the other elements of his genius to life for his various markets. He trained his first and second circle teams to present ideas, both internally and externally, in a compelling, memorable, and audience-focused manner. Ultimately, his entire cast served the establishment of an unforgettable, market-moving brand.

PART THREE

EXPANDING INNOVATION LITERACY

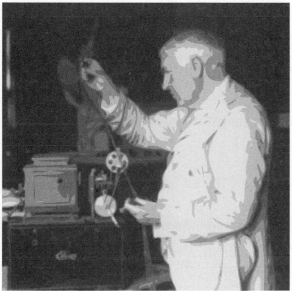

Chapter Eight

EDISON'S LEGACY IN THE TWENTY-FIRST CENTURY

Thomas Edison invented systematic innovation. His "invention factory" showed the world how to apply the processes and culture of innovation to create unprecedented value. More than his lighting system, phonograph, and kinetoscope, his method represents his greatest legacy.

Edison viewed innovation as a means to generate successful businesses, and he saw it as a powerful, positive social force. He invited scientists and inventors from all over the world to visit him at Menlo Park and West Orange, inspiring them to model their laboratories after his own. He funded the journal *Science* to ensure that scientific and innovation breakthroughs would be shared widely. He used emerging forms of media, including film—which he helped develop—to broadcast messages about his views of the future, generating pride in technological advancement. Edison wanted his innovations to benefit the entire world, but he believed that his country had a special role to play in inspiring human progress.

He appreciated that America's democratic institutions and culture of freedom provided the perfect setting for his accomplishments, and he sought to bring to fulfillment the potential championed by his hero, Abraham Lincoln. Lincoln believed that improvement in the quality of life would come through "discoveries and inventions," and that these were more likely to be made in an environment of freedom and opportunity. Lincoln gave his life to help make these ideals a possibility, and Edison dedicated his life to their realization.

These American ideals are manifest in the approach to creative problem solving known as "Yankee ingenuity." It was launched by Franklin, supported by Lincoln, and embodied supremely by Edison. This optimistic, creative, persistent, and practical approach to making a better world lives deep within the American soul. As F. Duane Ackerman, chairman and CEO of BellSouth and co-chairman

of the 2005 National Innovation Initiative stated, "If America were a company, freedom and exploration would be our core competencies."

But, as Sam Palmisano, chairman and CEO of IBM, and co-chairman of the National Innovation Initiative commented in a recent presentation, America is "somehow losing its edge" just as the world is "becoming dramatically more competitive." Barriers to innovation are falling in a "flat world" that places less emphasis on geography or bricks and mortar, and more emphasis on intellectual capital. Many American schools, modeled upon schedules and curricula developed in the Industrial Age, are losing their edge in developing intellectual capital, especially in math and engineering. And now, more than ever, it's important to think in global, creative, and cross-disciplinary terms. Innovation is occurring where cultures and disciplines intersect, and most American schools and businesses are not yet designed accordingly.

Just as Edison created a systematic approach to innovation, we must develop a systematic approach to teaching the skills and attitudes that will help us regain our competitive edge. As Gary Hamel states, "I don't think there are very many companies today that are managing their imagination capital or their entrepreneurial capital in any kind of systematic way, but that's where the competitive advantage comes from." Our schools, companies, and other institutions must reorganize themselves to develop these great, underutilized national resources. Imagination plus entrepreneurship equals innovation, and innovation equals competitive advantage.

But, before we can change our schools, companies, and institutions, we must change ourselves. The solution to Information Age problems isn't to be found in Industrial Age thinking and attitudes. Many well-meaning teachers, managers, and leaders are still approaching the business of innovation by doing the opposite of everything we've covered in these pages. A problem-centered mindset, linear thinking, grim effort, stove piping, and molecule fondling will not suffice. More energy invested into old habits isn't the answer. If Edison were alive now he'd be thrilled with the unprecedented opportunities available for learning, growth, and global business success. His legacy comes to life as you develop your own innovation literacy. Begin by embodying the five competencies for *Innovating Like Edison*. As you cultivate the attitudes and skills expressed in the twenty-five elements, you'll become a beacon of inspiration to those around you. We will help you consolidate everything you've learned and develop a plan for taking it further in the final chapter: the Edison Innovation Literacy Blueprint.

Chapter Nine

THE EDISON INNOVATION LITERACY BLUEPRINT

The Edison Innovation Literacy Blueprint will help you assess your current level of innovation literacy and make it easier for you to set goals and chart your progress. You will have the opportunity to rate yourself in each of the five competencies of innovation, on an element-by-element basis. Once you've completed this process for yourself, we will provide you with resources to do the same thing for your team—and eventually your whole organization.

Using the Edison Innovation Literacy Blueprint as a guide, you can begin mapping where you need to improve most, and create an action plan to *Innovate Like Edison*. Building your innovation literacy is a three-step process, as follows:

Step 1: Complete the assessments and scoring gauges for the five competencies of innovation.

Step 2: Chart your innovation literacy strengths and gaps on the Edison Innovation Literacy Blueprint.

Step 3: Make an individual innovation literacy development plan for yourself.

On the Blueprint included toward the end of this chapter (page 261), you will note that the five competencies of innovation are listed horizontally along the bottom of the chart, and the elements rise vertically above them.

Step 1 begins with the completion of the five twenty-question assessments included on the following pages, with one assessment per competency. Immediately following each assessment is a scoring gauge tailored to that specific competency. Read the instructions shown on each chart, transferring your point scores from the assessment onto the scoring gauge. Every scoring gauge is designed like a

229

radar screen. Your innovation strength areas lie closer to the outside of the gauge, and gap areas lie closer to the interior. A sample of a completed scoring gauge is included to give you a visual reference on how to complete it. Once you have completed all five assessments, you can summarize your scores by transferring them onto the master scoring gauge for elements, then the master scoring gauge for competencies, following the directions provided.

If you do not wish to make marks in your book as you work through the assessments and scoring gauges, or if you are away from your book and still want to take an assessment, you can download copies of the assessments and scoring gauges free at *www.innovatelikeedison.com*. Now, grab a pencil and complete Step 1 to begin identifying your current innovation literacy level.

Step 2 begins on page 255, once you have completed your assessment.

Assessment for Competency #1— Solution-centered Mindset

Please respond to the statements in these assessments by circling the choice that best describes your *current habits*, not your desired habits. Add the total points for each individual element on the "subtotal" lines shown. When you have completed the entire assessment, find the scoring gauge for Competency #1 and enter the point subtotals for each element, following the directions on the scoring gauge.

If you are completing the assessment based on either a team or division approach, neither category should exceed 50 (fifty) persons. If the group you desire to assess has more than 50 persons, separate it into smaller segments, then add your results together. This allows you to preserve valuable detail for each member of the group.

If the body of individuals you desire to assess does not lie within your company, or does not have a particular classification, please identify it as a "unit." When you enter your data on the scoring gauge chart, you will see a place to name your unit so you can recall whom you are scoring.

COMPETENCY #1: ASSESSMENT

I am completing this assessment based on (select one):

_____ Myself as an individual _____ My team _____ My division _____ My unit

Today's Date: _____

	Almost always true of me	Mostly true of me	Some-times true of me	Rarely true of me	Almost never true of me	NA/ Don't know
I write down my goals.	5	4	3	2	1	0
I review my written goals on a regular basis.	5	4	3	2	1	0
I visualize each of my goals, seeing them as completed.	5	4	3	2	1	0
I link a feeling of satisfaction, happiness, joy, or other positive emotions with each goal I set.	5	4	3	2	1	0

Element 1 subtotal: ____

I am optimistic and upbeat.	5	4	3	2	1	0
When I experience setbacks, I respond by coaching myself in a positive and adaptive manner.	5	4	3	2	1	0
I am always looking for solutions no matter how daunting the challenge.	5	4	3	2	1	0
I am able to inspire and influence others using positive persuasion rather than bullying or intimidation.	5	4	3	2	1	0

Element 2 subtotal: ____

	Almost always true of me	Mostly true of me	Some-times true of me	Rarely true of me	Almost never true of me	NA/ Don't know
I have a deep curiosity about many areas of life.	5	4	3	2	1	0
I believe learning continues throughout life, and doesn't end when formal education ends.	5	4	3	2	1	0
I read on a daily basis (books, periodicals, journals, etc.).	5	4	3	2	1	0
I am informed about current events.	5	4	3	2	1	0

Element 3 subtotal: ____

I view living as a series of experiments.	5	4	3	2	1	0
I seek to find practical ways to test the validity of my ideas.	5	4	3	2	1	0
I pursue solutions with relentless vigor and resolve.	5	4	3	2	1	0
I understand and can apply the scientific method.	5	4	3	2	1	0

Element 4 subtotal: ____

I am open-minded.	5	4	3	2	1	0
I seek out and empathically consider different perspectives besides my own.	5	4	3	2	1	0
I can separate my emotions from my analysis when evaluating a problem.	5	4	3	2	1	0
I do not discard unexpected out-comes or anomalies.	5	4	3	2	1	0

Element 5 subtotal: ____

Record your points for each element subtotal on the Competency #1 Scoring Gauge.

POINT SCORES FOR ELEMENTS 1–5

COMPETENCY #1—Scoring Gauge

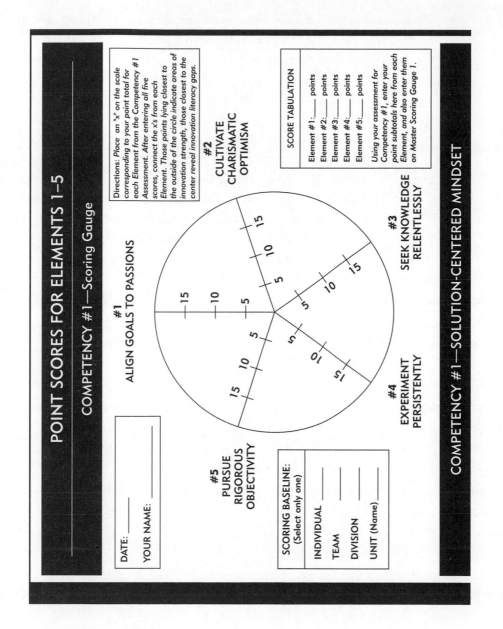

#1
ALIGN GOALS TO PASSIONS

#2
CULTIVATE
CHARISMATIC
OPTIMISM

#3
SEEK KNOWLEDGE
RELENTLESSLY

#4
EXPERIMENT
PERSISTENTLY

#5
PURSUE
RIGOROUS
OBJECTIVITY

Directions: *Place an "x" on the scale corresponding to your point total for each Element from the Competency #1 Assessment. After entering all five scores, connect the x's from each Element. Those points lying closest to the outside of the circle indicate areas of innovation strength, those closest to the center reveal innovation literacy gaps.*

SCORE TABULATION

Element #1: _____ points
Element #2: _____ points
Element #3: _____ points
Element #4: _____ points
Element #5: _____ points

Using your assessment for Competency #1, enter your point subtotals here from each Element, and also enter them on Master Scoring Gauge 1.

DATE: _____

YOUR NAME: _____

SCORING BASELINE:
(Select only one)

INDIVIDUAL _____
TEAM _____
DIVISION _____
UNIT (Name) _____

COMPETENCY #1—SOLUTION-CENTERED MINDSET

Assessment for Competency #2—
Kaleidoscopic Thinking

Please respond to the statements on the chart by circling the choice that best describes your *current habits*, not your desired habits. Add together the total points for each individual element on the subtotal lines shown. When you have completed the entire assessment, find the scoring gauge for Competency #2 and enter the point subtotals for each element, following the directions on the scoring gauge.

If you are completing the assessment based on either a team or division approach, neither category should exceed 50 (fifty) persons. If the group you desire to assess has more than 50 persons, separate it into smaller segments, then add your results together. This allows you to preserve valuable detail for each member of the group.

If the body of individuals you desire to assess does not lie within your company, or does not have a particular classification, please identify it as a "unit." When you enter your data on the scoring gauge chart, you will see a place to name your unit so you can recall whom you are scoring.

COMPETENCY #2: ASSESSMENT

I am completing this assessment based on (select one):

_____ Myself as an individual _____ My team _____ My division _____ My unit

Today's date: _____

	Almost always true of me	Mostly true of me	Some-times true of me	Rarely true of me	Almost never true of me	NA/ Don't know
I maintain a notebook.	5	4	3	2	1	0
I make entries in my notebook on a daily basis.	5	4	3	2	1	0
I write freely in my notebook, without censoring or editing.	5	4	3	2	1	0
I review my notebook entries on a regular basis.	5	4	3	2	1	0

Element 6 subtotal: ____

I can easily generate a lot of ideas.	5	4	3	2	1	0
I am skilled at making analogies.	5	4	3	2	1	0
I can readily engage my imagination by making up fantastical stories and image streams.	5	4	3	2	1	0
I can let my ideas flow without pre-judging, editing, or censoring them.	5	4	3	2	1	0

Element 7 subtotal: ____

I trust my hunches and intuitions.	5	4	3	2	1	0
I can translate ideas or solutions from one project to another.	5	4	3	2	1	0
I am able to deal with information gaps or ambiguities in a productive way.	5	4	3	2	1	0

	Almost always true of me	Mostly true of me	Some-times true of me	Rarely true of me	Almost never true of me	NA/ Don't know
I am able to see the big picture as well as the details.	5	4	3	2	1	0

Element 8 subtotal: ____

I can picture my ideas in my mind's eye.	5	4	3	2	1	0
I use drawing, sketching, and creative doodling as part of my problem-solving process.	5	4	3	2	1	0
I use models or graphical simulations of my ideas to expand my thinking.	5	4	3	2	1	0
I use pictures, sketches, or creative doodles to help share my ideas with others.	5	4	3	2	1	0

Element 9 subtotal: ____

I regularly question conventional wisdom.	5	4	3	2	1	0
I regularly question my own assumptions.	5	4	3	2	1	0
I am able to stand my ground when everyone around me disagrees with me.	5	4	3	2	1	0
I am able to be independent-minded without being pigheaded.	5	4	3	2	1	0

Element 10 subtotal: ____

Record your points for each element subtotal on the Competency #2 Scoring Gauge.

POINT SCORES FOR ELEMENTS 6–10

COMPETENCY #2—Scoring Gauge

Directions: Place an "x" on the scale corresponding to your point total for each Element from the Competency #2 Assessment. After entering all five scores, connect the x's from each Element. Those points lying closest to the outside of the circle indicate areas of innovation strength, those closest to the center reveal innovation literacy gaps.

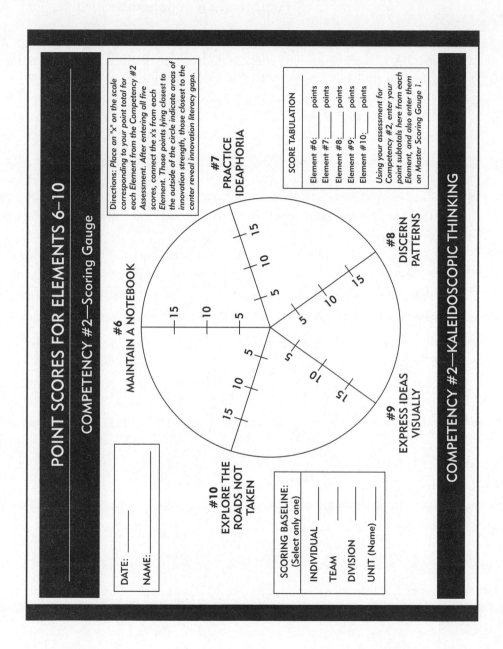

#6
MAINTAIN A NOTEBOOK

#7
PRACTICE
IDEAPHORIA

#8
DISCERN
PATTERNS

#9
EXPRESS IDEAS
VISUALLY

#10
EXPLORE THE
ROADS NOT
TAKEN

SCORE TABULATION

Element #6: _____ points
Element #7: _____ points
Element #8: _____ points
Element #9: _____ points
Element #10: _____ points

Using your assessment for Competency #2, enter your point subtotals here from each Element, and also enter them on Master Scoring Gauge 1.

DATE: _____

NAME: _____

SCORING BASELINE:
(Select only one)

INDIVIDUAL _____
TEAM _____
DIVISION _____
UNIT (Name) _____

COMPETENCY #2—KALEIDOSCOPIC THINKING

Assessment for Competency #3— Full-spectrum Engagement

Please respond to the statements on the chart by circling the choice that best describes your *current habits*, not your desired habits. Add together the total points for each individual element on the subtotal lines shown. When you have completed the entire assessment, find the scoring gauge for Competency #3 and enter the point subtotals for each element, following the directions on the scoring gauge.

If you are completing the assessment based on either a team or division approach, neither category should exceed 50 (fifty) persons. If the group you desire to assess has more than 50 persons, separate it into smaller segments, then add your results together. This allows you to preserve valuable detail for each member of the group.

If the body of individuals you desire to assess does not lie within your company, or does not have a particular classification, please identify it as a "unit." When you enter your data on the scoring gauge chart, you will see a place to name your unit so you can recall whom you are scoring.

COMPETENCY #3: ASSESSMENT

I am completing this assessment based on (select one):

_____ Myself as an individual _____ My team _____ My division _____ My unit

Today's date: _____

	Almost always true of me	Mostly true of me	Some-times true of me	Rarely true of me	Almost never true of me	NA/ Don't know
I use breaks to optimize my energy and productivity.	5	4	3	2	1	0
I know how and when to shift from one topic to another to optimize my energy and productivity.	5	4	3	2	1	0
I experience working in a flow state on a regular basis.	5	4	3	2	1	0
When I'm running a meeting, I use breaks and topic shifts to optimize the group's energy and productivity.	5	4	3	2	1	0

Element 11 subtotal: _____

I value playfulness as a quality that gives me access to new ways of thinking and being.	5	4	3	2	1	0
I am able to be lighthearted in the midst of stressful situations.	5	4	3	2	1	0
I can laugh at myself.	5	4	3	2	1	0
I use humor to help others relax and refocus.	5	4	3	2	1	0

Element 12 subtotal: _____

I understand the concept of intellectual capital.	5	4	3	2	1	0
I know the difference between a patent, a trademark, a service mark, and a trade secret.	5	4	3	2	1	0

	Almost always true of me	Mostly true of me	Some-times true of me	Rarely true of me	Almost never true of me	NA/ Don't know
I am able to share information about my projects in a way that does not compromise proprietary methods or processes.	5	4	3	2	1	0
I appreciate the role of protecting intellectual property in the process of innovation.	5	4	3	2	1	0

Element 13 subtotal: ____

I am skilled at eliminating the unnecessary.	5	4	3	2	1	0
I am able to give clear, concise directions.	5	4	3	2	1	0
I can remain calm and centered while embracing a wide array of challenges.	5	4	3	2	1	0
In the midst of complex situations, I'm able to clearly prioritize a way forward.	5	4	3	2	1	0

Element 14 subtotal: ____

I spend some time in solitude each day.	5	4	3	2	1	0
I have a favorite place where I can go to spend time in quiet contemplation.	5	4	3	2	1	0
I have practical strategies for finding peace in the midst of a chaotic environment.	5	4	3	2	1	0
I value the way my time alone prepares me to be more sensitive and attuned to others.	5	4	3	2	1	0

Element 15 subtotal: ____

Record your points for each element subtotal on the Competency #3 Scoring Gauge.

POINT SCORES FOR ELEMENTS 11–15

COMPETENCY #3—Scoring Gauge

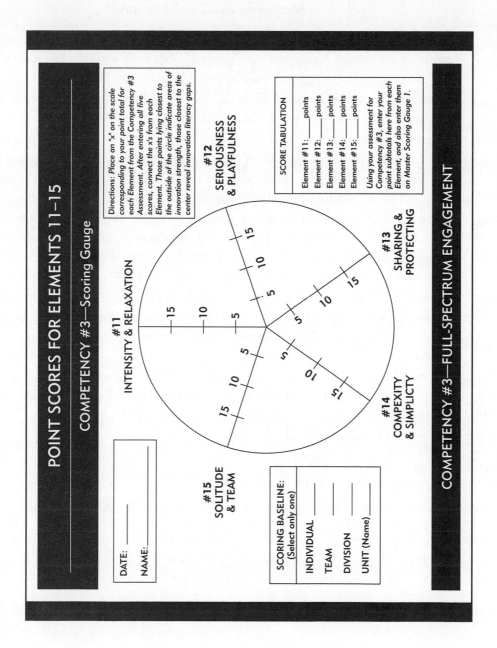

Directions: Place an "x" on the scale corresponding to your point total for each Element from the Competency #3 Assessment. After entering all five scores, connect the x's from each Element. Those points lying closest to the outside of the circle indicate areas of innovation strength, those closest to the center reveal innovation literacy gaps.

#11 INTENSITY & RELAXATION

#12 SERIOUSNESS & PLAYFULNESS

#13 SHARING & PROTECTING

#14 COMPEXITY & SIMPLICTY

#15 SOLITUDE & TEAM

SCORE TABULATION

Element #11: _____ points
Element #12: _____ points
Element #13: _____ points
Element #14: _____ points
Element #15: _____ points

Using your assessment for Competency #3, enter your point subtotals here from each Element, and also enter them on Master Scoring Gauge 1.

DATE: _____

NAME: _____

SCORING BASELINE:
(Select only one)

INDIVIDUAL _____
TEAM _____
DIVISION _____
UNIT (Name) _____

COMPETENCY #3—FULL-SPECTRUM ENGAGEMENT

Assessment for Competency #4—
Master-mind Collaboration

Please respond to the statements on the chart by circling the choice that best describes your *current habits*, not your desired habits. Add together the total points for each individual element on the subtotal lines shown. When you have completed the entire assessment, find the scoring gauge for Competency #4 and enter the point subtotals for each element, following the directions on the scoring gauge.

Please note that this assessment as well as the assessment for Competency #5 have added the phrase "True of me or my environment." Many of the statements you will be responding to address your workplace or work environment. If you have multiple work environments—such as multiple divisional locations—or if you are a municipal body that is part of a state or county, for example—name the specific environment you are assessing on the line marked "my environment."

If you are completing the assessment based on either a team or division approach, neither category should exceed 50 (fifty) persons. If the group you desire to assess has more than 50 persons, separate it into smaller segments, then add your results together. This allows you to preserve valuable detail for each member of the group.

If the body of individuals you desire to assess does not lie within your company, or does not have a particular classification, please identify it as a "unit." When you enter your data on the scoring gauge chart, you will see a place to name your unit so you can recall whom you are scoring.

COMPETENCY #4: ASSESSMENT

I am completing this assessment based on (select one):

_____ Myself as an individual _____ My team _____ My division _____ My unit

Today's date _____

My environment _____

	Almost always true of me or my environment	Mostly true of me or my environment	Some-times true of me or my environment	Rarely true of me or my environment	Almost never true of me or my environment	NA/ Don't know
When I'm evaluating candidates for employment, I pose questions to determine how well they "think on their feet."	5	4	3	2	1	0
When I'm evaluating candidates for employment, I assess their breadth of skills and interests as well as their areas of expertise.	5	4	3	2	1	0
When I'm evaluating candidates for employment, I place them in contexts similar to the ones they would experience as actual employees.	5	4	3	2	1	0
When I'm evaluating candidates for employment, I assess how well they would integrate with my team.	5	4	3	2	1	0

Element 16 subtotal: _____

	Almost always true of me or my environment	Mostly true of me or my environment	Some-times true of me or my environment	Rarely true of me or my environment	Almost never true of me or my environment	NA/ Don't know
I value a multidisciplinary approach to problem solving.	5	4	3	2	1	0
I regularly seek input from people with different backgrounds and perspectives.	5	4	3	2	1	0
I respect and encourage diverse approaches to accomplishing an objective.	5	4	3	2	1	0
I am aware of different learning styles and personality types, and leverage these differences for optimal results.	5	4	3	2	1	0

Element 17 subtotal: ____

I am aware of the ways in which fear prevents openness in my organization.	5	4	3	2	1	0
I actively seek new and creative ways to encourage an open exchange of ideas.	5	4	3	2	1	0
I use open-ended questions to encourage people around me to share ideas freely.	5	4	3	2	1	0
I encourage others to step outside of mainstream thought.	5	4	3	2	1	0

Element 18 subtotal: ____

	Almost always true of me or my environment	Mostly true of me or my environment	Some-times true of me or my environment	Rarely true of me or my environment	Almost never true of me or my environment	NA/ Don't know
I understand how to use incentives and compensation as rewards for collaboration.	5	4	3	2	1	0
I take the initiative to creatively reward collaborative efforts by my colleagues at all levels.	5	4	3	2	1	0
I strive to make the collaborative process in my organization intrinsically rewarding.	5	4	3	2	1	0
I am aware of the behaviors and practices that have discouraged collaboration in my organization.	5	4	3	2	1	0

Element 19 subtotal: _____

I have an accurate and up-to-date record of everyone in my network.	5	4	3	2	1	0
I regularly touch base with all the people in my network.	5	4	3	2	1	0
I target diverse resources and key influencers in my networking efforts.	5	4	3	2	1	0
In my networking efforts, I help others achieve their goals, while also focusing on the outcomes I seek.	5	4	3	2	1	0

Element 20 subtotal: _____

Record your points for each element subtotal on the Competency #4 Scoring Gauge.

POINT SCORES FOR ELEMENTS 16–20

COMPETENCY #4—Scoring Gauge

DATE: _____

NAME: _____

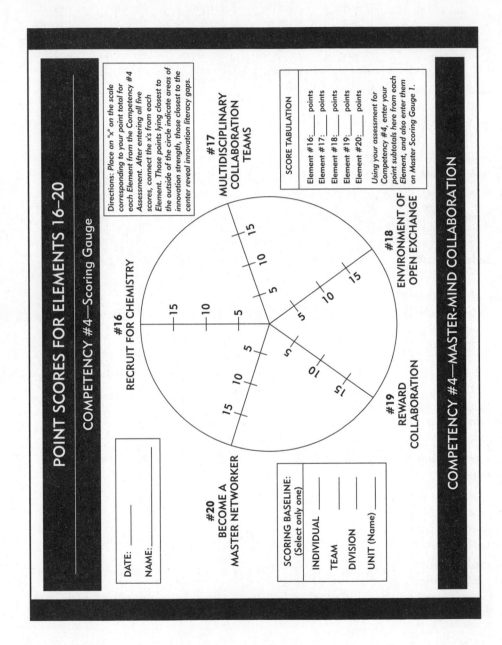

Directions: *Place an "x" on the scale corresponding to your point total for each Element from the Competency #4 Assessment. After entering all five scores, connect the x's from each Element. Those points lying closest to the outside of the circle indicate areas of innovation strength, those closest to the center reveal innovation literacy gaps.*

#17
MULTIDISCIPLINARY COLLABORATION TEAMS

#16
RECRUIT FOR CHEMISTRY

#18
ENVIRONMENT OF OPEN EXCHANGE

#19
REWARD COLLABORATION

#20
BECOME A MASTER NETWORKER

SCORE TABULATION

Element #16: _____ points
Element #17: _____ points
Element #18: _____ points
Element #19: _____ points
Element #20: _____ points

Using your assessment for Competency #4, enter your point subtotals here from each Element, and also enter them on Master Scoring Gauge 1.

SCORING BASELINE:
(Select only one)

INDIVIDUAL _____

TEAM _____

DIVISION _____

UNIT (Name) _____

COMPETENCY #4—MASTER-MIND COLLABORATION

Super-value Creation Assessment

Please respond to the statements on the chart by circling the choice that best describes your *current habits*, not your desired habits. Add together the total points for each individual element on the subtotal lines shown. When you have completed the entire assessment, find the scoring gauge for Competency #5 on the page immediately following the assessment, and enter the point subtotals for each element as directed on the scoring gauge.

Please note that this assessment includes the phrase "True of me or my environment." Many of the statements you will be responding to address your workplace or work environment. If you have multiple work environments—such as multiple divisional locations—or if you are a municipal body that is part of a state or county, for example—name the specific environment you are assessing on the line marked "my environment."

If you are completing the assessment based on either a team or division approach, neither category should exceed 50 (fifty) persons.

If the body of individuals you desire to assess does not lie within your company, or does not have a particular classification, please identify it as a "unit." When you enter your data on the scoring gauge chart, you will see a place to name your unit so you can recall whom you are scoring.

COMPETENCY #5: ASSESSMENT

I am completing this assessment based on (select one):

_____ Myself as an individual _____ My team _____ My division _____ My unit

Today's date _____

My environment _____

	Almost always true of me or my environment	Mostly true of me or my environment	Some-times true of me or my environment	Rarely true of me or my environment	Almost never true of me or my environment	NA/ Don't know
I seek to identify trends.	5	4	3	2	1	0
I regularly search for quality, pricing, technology, or other kinds of gaps in the marketplace.	5	4	3	2	1	0
I am aware of my organization's core strengths.	5	4	3	2	1	0
I seek to make practical linkages between the trends I observe, the gaps I identify, and my organization's core strengths.	5	4	3	2	1	0

Element 21 subtotal: ____

I am focused on the needs of my customers/ clients.	5	4	3	2	1	0

	Almost always true of me or my environment	Mostly true of me or my environment	Some-times true of me or my environment	Rarely true of me or my environment	Almost never true of me or my environment	NA/ Don't know
I take creative initiative to find out more about my customers'/clients' needs.	5	4	3	2	1	0
When I learn about a challenge my customer/ client has with one of my products/services, I take immediate action to address it.	5	4	3	2	1	0
I can consistently translate my areas of expertise in terms my customers/clients can understand.	5	4	3	2	1	0

Element 22 subtotal: _____

	Almost always	Mostly	Some-times	Rarely	Almost never	NA/ Don't know
I know what a business model is.	5	4	3	2	1	0
I am aware of different types of business mod-els, and the importance of finding the right one.	5	4	3	2	1	0
I am aware of the importance of modifying business models in response to changing market conditions or technology shifts.	5	4	3	2	1	0
My company's business model is optimized to drive innovation.	5	4	3	2	1	0

Element 23 subtotal: _____

	Almost always true of me or my environment	Mostly true of me or my environment	Some-times true of me or my environment	Rarely true of me or my environment	Almost never true of me or my environment	NA/ Don't know
I appreciate the importance of solving a problem when it's small.	5	4	3	2	1	0
Before my team, group, or organization introduces something new to the marketplace, we make a quantitative assessment of anticipated costs, profits, and market size.	5	4	3	2	1	0
Before launching a new initiative, my team, group, or organization sets a projected budget and timetable for the scale-up process.	5	4	3	2	1	0
Before launching an innovation, my team, group, or organization connects the team(s) involved in the origination of the new product/service with the team(s) involved in the scale-up early in the process.	5	4	3	2	1	0

Element 24 subtotal: ____

	Almost always true of me or my environment	Mostly true of me or my environment	Some-times true of me or my environment	Rarely true of me or my environment	Almost never true of me or my environment	NA/ Don't know
I am committed to a process of continuously improving the effectiveness of my communication.	5	4	3	2	1	0
When preparing a presentation, I focus on ensuring that my audience will remember my message in addition to understanding it.	5	4	3	2	1	0
In all my communications, I measure my success by the response I get from my audience.	5	4	3	2	1	0
I recognize myself as a brand whose distinctive strengths I must nurture over the long term.	5	4	3	2	1	0

Element 25 subtotal: _____

Record your points for each element subtotal on the Competency #5 Scoring Gauge.

POINT SCORES FOR ELEMENTS 21–25

COMPETENCY #5—Scoring Gauge

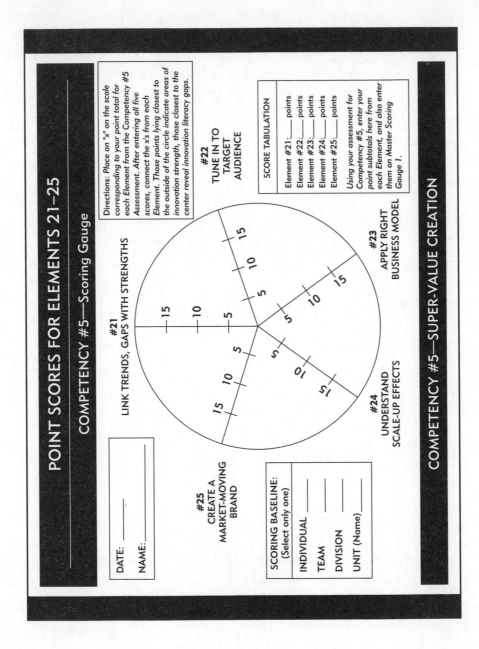

Directions: *Place an 'x' on the scale corresponding to your point total for each Element from the Competency #5 Assessment. After entering all five scores, connect the x's from each Element. Those points lying closest to the outside of the circle indicate areas of innovation strength, those closest to the center reveal innovation literacy gaps.*

#22
TUNE IN TO TARGET AUDIENCE

#21
LINK TRENDS, GAPS WITH STRENGTHS

15
10
5

15
10
5

5
10
15

5
10
15

#23
APPLY RIGHT BUSINESS MODEL

#24
UNDERSTAND SCALE-UP EFFECTS

#25
CREATE A MARKET-MOVING BRAND

SCORE TABULATION

Element #21: _____ points
Element #22: _____ points
Element #23: _____ points
Element #24: _____ points
Element #25: _____ points

Using your assessment for Competency #5, enter your point subtotals here from each Element, and also enter them on Master Scoring Gauge 1.

DATE: _____

NAME: _____

SCORING BASELINE:
(Select only one)

INDIVIDUAL _____
TEAM _____
DIVISION _____
UNIT (Name) _____

COMPETENCY #5—SUPER-VALUE CREATION

MASTER SCORING GAUGE 1—ELEMENTS

ELEMENTS 1–25: Translating "Points" to "Percentages"

DATE: _____ NAME: _____

Directions:

This chart translates your individual point subtotals for each Element into a percentage. Percentages are useful when summarizing your individual scores with your team's scores.

Begin by recording your Element subtotals for all 25 Elements in the boxes noted. Then, using the chart at right, translate point subtotals into percentages.

Add the percentages for each Element, creating a "Competency Total %," then divide by 5. If Element scores fall below 10 points, translate these as "0%." Enter each of your five Competency Total %'s on Master Scoring Gauge 2. This will help you visually "see" your innovation literacy scores in graphical form.

You can use the scoring box in the lower right corner to summarize individual or team percentages by Competency.

TRANSLATING POINTS INTO PERCENTAGES

Points		%
10	=	50%
11	=	55%
12	=	60%
13	=	65%
14	=	70%
15	=	75%
16	=	80%
17	=	85%
18	=	90%
19	=	95%
20	=	100%

POINTS TO PERCENTAGES: COMPETENCY #1:

	Points		%
Element 1:	___	=	___
Element 2:	___	=	___
Element 3:	___	=	___
Element 4:	___	=	___
Element 5:	___	=	___

Comp 1 Total % ___ ÷ 5 = ___
Record total on Master Scoring Gauge 2

POINTS TO PERCENTAGES: COMPETENCY #2:

	Points		%
Element 6:	___	=	___
Element 7:	___	=	___
Element 8:	___	=	___
Element 9:	___	=	___
Element 10:	___	=	___

Comp 2 Total % ___ ÷ 5 = ___
Record total on Master Scoring Gauge 2

POINTS TO PERCENTAGES: COMPETENCY #3:

	Points		%
Element 11:	___	=	___
Element 12:	___	=	___
Element 13:	___	=	___
Element 14:	___	=	___
Element 15:	___	=	___

Comp 3 Total % ___ ÷ 5 = ___
Record total on Master Scoring Gauge 2

POINTS TO PERCENTAGES: COMPETENCY #4:

	Points		%
Element 16:	___	=	___
Element 17:	___	=	___
Element 18:	___	=	___
Element 19:	___	=	___
Element 20:	___	=	___

Comp 4 Total % ___ ÷ 5 = ___
Record total on Master Scoring Gauge 2

POINTS TO PERCENTAGES: COMPETENCY #5:

	Points		%
Element 21:	___	=	___
Element 22:	___	=	___
Element 23:	___	=	___
Element 24:	___	=	___
Element 25:	___	=	___

Comp 5 Total % ___ ÷ 5 = ___
Record total on Master Scoring Gauge 2

CALCULATING INDIVIDUAL OR TEAM COMPETENCY PERCENTAGES

		%
Comp 1:		%
Comp 2:		%
Comp 3:		%
Comp 4:		%
Comp 5:		%

Calculate your team percentage by adding individual Competency percentage scores together for each team member, then divide by the total number of members on the team.

ESTABLISHIING INNOVATION LITERACY

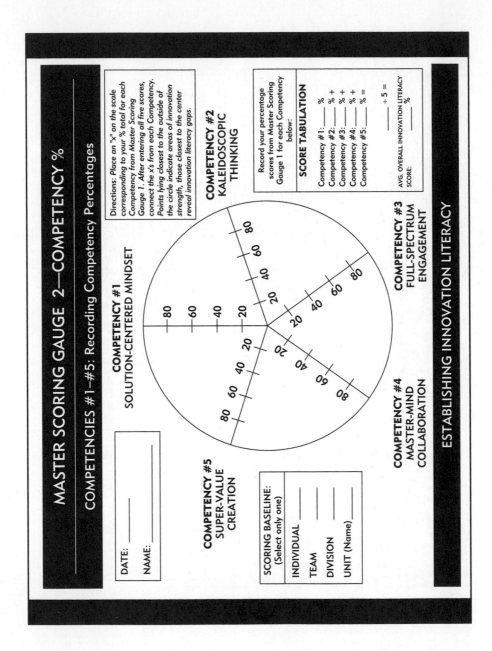

MASTER SCORING GAUGE 2—COMPETENCY %

COMPETENCIES #1–#5: Recording Competency Percentages

DATE: _____

NAME: _____

Directions: *Place an "x" on the scale corresponding to your % total for each Competency from Master Scoring Gauge 1. After entering all five scores, connect the x's from each Competency. Points lying closest to the outside of the circle indicate areas of innovation strength, those closest to the center reveal innovation literacy gaps.*

COMPETENCY #1
SOLUTION-CENTERED MINDSET

COMPETENCY #2
KALEIDOSCOPIC THINKING

COMPETENCY #3
FULL-SPECTRUM ENGAGEMENT

COMPETENCY #4
MASTER-MIND COLLABORATION

COMPETENCY #5
SUPER-VALUE CREATION

Record your percentage scores from Master Scoring Gauge 1 for each Competency below:

SCORE TABULATION

Competency #1: _____ %
Competency #2: _____ % +
Competency #3: _____ % +
Competency #4: _____ % +
Competency #5: _____ % =

_____ ÷ 5 =

AVG. OVERALL INNOVATION LITERACY
SCORE: _____ %

SCORING BASELINE:
(Select only one)

INDIVIDUAL _____

TEAM _____

DIVISION _____

UNIT (Name) _____

ESTABLISHING INNOVATION LITERACY

Congratulations! You have now completed Step 1 of improving your innovation literacy, and learning to *Innovate Like Edison*. Next, you will map your innovation literacy profile using results from the scoring gauges. If you have downloaded the charts from the *www.innovatelikeedison.com* Web site, gather your completed scoring gauge charts. As well, find orange, yellow, green, blue, and purple colored markers or colored pencils, and locate the Innovation Literacy Blueprint on page 261.

Look specifically at the scoring gauge charts for each competency, and locate your individual element scores. Now, find the elements for which you have the highest point totals. Consulting the table below, see if any of your element scores were at 16 points or more. If so, find the squares on the Edison Innovation Blueprint on page 261 that correspond to these scores, and color them purple. You've achieved a "Superior" innovation literacy rating for these elements. Now, find the next highest element scores. See if you have any with at least a 15-point rating. If so, find the squares on the Edison Innovation Blueprint which correspond to these scores and color them blue. For these elements you have achieved an "Excellent" rating. Continue this same process until you have found and colored in the element squares for scores of at least 14 points (Above Average), 12 points (Average), and 10 points (Below Average). Here is a summary of the innovation literacy rating levels:

Innovation Literacy Tracking Chart—Individual Basis

Innovation Literacy Rating Level	Total Number of Elements Activated	Innovation Literacy Rating Color
Below Average	Elements with at least 10–11 points each	Orange
Average	Elements with at least 12–13 points each	Yellow
Above Average	Elements with at least 14 points each	Green
Excellent	Elements with at least 15 points each	Blue
Superior	Elements with at least 16 points each or higher	Purple

Once you've finished this part of Step 2, you may see that some of your Edison Innovation Literacy Blueprint squares on page 261 have no color in them at all. These squares are "inactive." You have not yet achieved a threshold level of innovation literacy for these elements because your score is under 10 points. Ten points is the threshold because it represents 50 percent of the 20 points possible for each element. Don't worry about the inactive elements on your Innovation Blueprint. Inactive elements will be a great place to begin building your innovation literacy in Step 3.

Now, spend several minutes looking at your Blueprint with all the squares you've just colored in, and find the colored squares with elements corresponding to your highest scores. These are your "leverage" points. You will want to use these high scores as "levers" to begin raising lower scores in other elements, or to bridge into new territory on the Blueprint where you have little or no innovation literacy currently established.

To get you started, here are sample results from Pat, one of our prepublication readers, showing scores from Competency #1 plus a 90-day plan to build innovation literacy. Based on Pat's scores, shown on page 257, Pat made a notebook entry using the SMART EDISON approach outlined in Element 1. Pat developed a step-by-step plan to move forward by leveraging high scores to pull up low scores en route to completing a goal of becoming a vice president in two years. You can apply the same philosophy. Although it's possible to improve your innovation literacy in multiple competencies at the same time, we suggest you begin by focusing on just one.

Here is Pat's 90-day notebook entry:

Monday

Yesterday, I took the hundred-question innovation literacy assessment. I noticed some of my lowest scores were in Competency #1, so I picked this as my first priority. If I want to be a division vice president, I must improve my solution orientation. Here are my scores:

Element 1 (goals): 8 points=Below Average

Element 2 (optimism): 19 points=Superior

Element 3 (knowledge): 15 points=Excellent

Element 4 (experimentation): 14 points=Above Average

Element 5 (objectivity): 10 points=Average

POINT SCORES FOR ELEMENTS 1–5

COMPETENCY #1—Scoring Gauge

DATE: 10-28-07

YOUR NAME: PAT

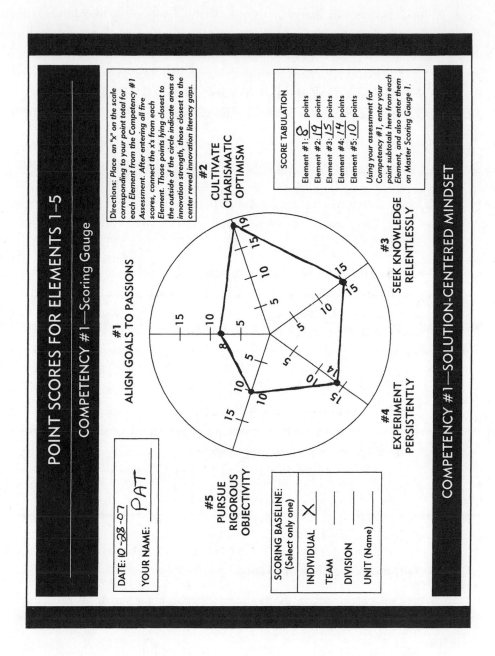

Directions: Place an "x" on the scale corresponding to your point total for each Element from the Competency #1 Assessment. After entering all five scores, connect the x's from each Element. Those points lying closest to the outside of the circle indicate areas of innovation strength, those closest to the center reveal innovation literacy gaps.

#1
ALIGN GOALS TO PASSIONS

#2
CULTIVATE CHARISMATIC OPTIMISM

#3
SEEK KNOWLEDGE RELENTLESSLY

#4
EXPERIMENT PERSISTENTLY

#5
PURSUE RIGOROUS OBJECTIVITY

SCORE TABULATION

Element #1: 8 points
Element #2: 19 points
Element #3: 15 points
Element #4: 14 points
Element #5: 10 points

Using your assessment for Competency #1, enter your point subtotals here from each Element, and also enter them on Master Scoring Gauge 1.

SCORING BASELINE:
(Select only one)

INDIVIDUAL X

TEAM _____

DIVISION _____

UNIT (Name) _____

COMPETENCY #1—SOLUTION-CENTERED MINDSET

I realize I can use my strong scores in "optimism" and "knowledge" to build up my low scores in "goals" and "objectivity." I can view the whole thing as an "experiment," and actually bring up my score up for that element as well.

Here is my plan for how to do this, using SMART EDISON:

MY NINETY-DAY INNOVATION LITERACY PLAN— COMPETENCY #1 SMART EDISON

S (Specific)—Raise my "goals" score to 15 or higher by writing down my goals. Raise my "objectivity" score to 14 or higher by seeking out perspectives beside my own. Reframing my outlook to a solution orientation rather than a worry orientation will release my constant concern about how everything will turn out. I can view outcomes with more "objectivity," knowing that I can always try again.

M (Measurable)—Review my written goals every week on Sundays. Retake the Competency #1 assessment on the last Sunday of the month for the next three months. Track results against my targets. At the end of ninety days I will complete the relevant charts and measure my progress.

A (Accountability)—I'm accountable! I will share my goals and results with my friend Chris, who has been a mentor for years.

R (Relevant)—It's relevant for me to focus on becoming more solution-oriented because I see solutions are valued in my organization. I also want to be a happier, more solution-centered person.

T (Timeline)—I'll complete my first ninety-day goal in time for our organization's tenth anniversary conference, where I can share results with Chris and other colleagues.

E (Emotion)—I feel excited, exuberant, and joyful, envisioning myself sharing my progress at the conference.

D (Decision)—I *am* solution-oriented and objective in my business life—and my personal life!

I (Integration)—I can connect my strong, optimistic attitude to my goal to become solution-oriented. I can start looking at what I read every day with an objective viewpoint rather than an emotional viewpoint. Overall, achieving my goal will help me become a more effective leader, and a more effective person.

S (Sensory)—I see myself sharing my results at the anniversary conference, beaming as I show my colleagues my Scoring Gauge charts, my notebook, and telling them how I am newly looking at my job—and my life. I see smiles on their faces, and hear them asking me how they can expand their innovation literacy, too.

O (Optimistic)—I am a solution-oriented person who suggests experiments on how to change things in my organization. I experiment in my life with ways to change things for the better.

N (Now)—I am *Innovating Like Edison* every day, manifesting my goals, working forward with optimism, and viewing outcomes objectively. For the first time, I see how each day can be a mini-experiment helping me move ahead productively.

Just as Pat has done, begin expanding your innovation literacy by first setting a big goal, then logging your first ninety-day SMART EDISON plan into your notebook. Be sure to cover all aspects of SMART EDISON in your plan. As you work through it, refer to the assessments in the book to see what specific skills you need to build upon. You can also refer to the competency chapters earlier in the book, and consult the resources we have included for your use in the Resources and Reference Notes (see page 263).

Additional copies of the Edison Innovation Literacy Blueprint, including assessments and scoring gauges, are available free of charge on our Web site *www.innovatelikeedison.com*. Your organization can begin creating a *corporate innovation infrastructure* by committing to train a critical mass of your people to become innovation literate. As innovation literacy expands within an organization, a culture of innovation emerges. As more individuals at all levels embody the competencies, your organization will develop a unique culture of innovation that cannot be copied by competitors. If you'd like to share the innovation literacy process with your innovation team and introduce it throughout your entire organization, we offer free guidelines to help you begin optimizing your organization for innovation, also available at *www.innovatelikeedison.com*.

Thomas Edison lives as a timeless example of the principles of innovation and success. As he reminds us, "The value of an idea lies in the using of it." Your life is your laboratory for exploring his practical wisdom. By committing yourself to the disciplines of innovation literacy, you will become a force for creative illumination in your daily life, at work, and at home.

THE EDISON INNOVATION LITERACY BLUEPRINT™
5 Competencies, 25 Elements

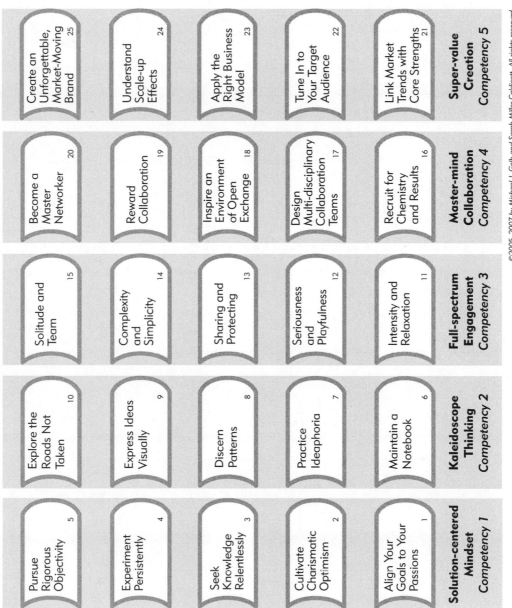

Solution-centered Mindset Competency 1	Kaleidoscope Thinking Competency 2	Full-spectrum Engagement Competency 3	Master-mind Collaboration Competency 4	Super-value Creation Competency 5
Pursue Rigorous Objectivity — 5	Explore the Roads Not Taken — 10	Solitude and Team — 15	Become a Master Networker — 20	Create an Unforgettable, Market-Moving Brand — 25
Experiment Persistently — 4	Express Ideas Visually — 9	Complexity and Simplicity — 14	Reward Collaboration — 19	Understand Scale-up Effects — 24
Seek Knowledge Relentlessly — 3	Discern Patterns — 8	Sharing and Protecting — 13	Inspire an Environment of Open Exchange — 18	Apply the Right Business Model — 23
Cultivate Charismatic Optimism — 2	Practice Ideaphoria — 7	Seriousness and Playfulness — 12	Design Multi-disciplinary Collaboration Teams — 17	Tune In to Your Target Audience — 22
Align Your Goals to Your Passions — 1	Maintain a Notebook — 6	Intensity and Relaxation — 11	Recruit for Chemistry and Results — 16	Link Market Trends with Core Strengths — 21

To aid you in visualizing Edison's Five Competencies of Innovation and the twenty-five elements that support them, we've created a summary chart called the Edison Innovation Literacy Blueprint. The Five Competencies of Innovation are listed horizontally along the bottom of the chart, and the elements rise vertically above them. Use the Edison Innovation Literacy Blueprint for reference as you begin *Innovating Like Edison*.

RESOURCES AND REFERENCE NOTES

RECOMMENDED GENERAL REFERENCES:

William Shakespeare: *The Oxford Shakespeare: The Complete Works*, edited by Stanley Wells, Gary Taylor, John Jowett, and William Montgomery, Oxford University Press, New York, 2005.

Napoleon Hill: *Think and Grow Rich!* Napoleon Hill, Aventine Press, www.aventinepress.com, 2004; *The Laws of Success*, Napoleon Hill, www.bnpublishing.com, 2006; contact the Napoleon Hill Foundation for more information about this seminal force in positive psychology: www.naphill.com.

Peter Drucker: *The Essential Drucker: The Best of Sixty Years of Peter Drucker's Essential Writings on Management*, Peter F. Drucker, Collins Business, New York, 2003; *Managing for the Future: The 1990s and Beyond*, Peter F. Drucker, Plume, New York, 1993; www.druckerarchives.net.

Ralph Waldo Emerson: *The Essential Writings of Ralph Waldo Emerson*, edited by Brooks Atkinson, Modern Library, New York, 2000; The Ralph Waldo Emerson Society: www.cas.sc.edu/engl/emerson.

Benjamin Franklin: *The First American: The Life and Times of Benjamin Franklin*, H. W. Brands, Anchor Books, New York, 2002; *Benjamin Franklin: An American Life*, Walter Isaaacson, Simon & Schuster, New York, 2004.

Michael Faraday: *The Electric Life of Michael Faraday*, Alan W. Hirshfeld, Walker and Co., New York, 2006.

Robert Ingersoll: *Best of Robert Ingersoll: Selection from His Writings and Speeches*, edited by Roger E. Greeley, Prometheus, Amherst, NY, 1993.

Abraham Lincoln: *Selected Speeches and Writings: Abraham Lincoln*, Vintage, 1992; Abraham Lincoln, Lord Charnwood, Dodo Press, New York, 2007.

Thomas Paine: *Thomas Paine: Collected Writings: Common Sense/The Crisis/Rights of Man/The Age of Reason/Pamphlets, Articles, and Letters*, edited by Eric Foner, Library of America, New York, 1995.

xix   Epigraph: "My philosophy of life is work . . ." Thomas A. Edison, *Edison & Ford Quote Book*, Edison & Ford Winter Estates, Fort Myers, FL, page 5.

CHAPTER 1—INTRODUCTION: TURNING ON THE LIGHT

3   "If we did all the things we are capable of . . ." Thomas A. Edison, *Edison & Ford Quote Book,* page 24.

3–4   This story is an adaptation taken from the memoirs of Edison's laboratory assistant Francis Jehl, *Menlo Park Reminiscences, Volume I*, World Scientific, Hackensack, NJ, 1990, Chapter XLVI, Oct 21, 1879, pp. 351–357. Jehl indicates that the "life test" experiment took place over a period of several days beginning October 19, 1879, although other laboratory records show it actually all took place on October 22. Notes kept by Charles Batchelor and Francis Upton also suggest greater involvement on their part than Jehl infers. Jehl and others citing his *Reminiscences* erroneously indicate the "life test" lasted for 45 hours; laboratory records confirm it was 14.5 hours, as noted in our opening story.

6   "Hell, there are no rules here . . ." Thomas A. Edison, *Edison & Ford Quote Book*, page 13.

7   *The World Is Flat: A Brief History of the 21st Century*, Thomas Friedman, Farrar, Straus and Giroux, New York, 2005. Friedman describes how an extraordinary confluence of social, political, economic, and technological forces are changing the balance of power in the world today, moving from an emphasis on bricks-and-mortar and hard assets to a "flat world," in which knowledge is the key asset, and where innovation is the key ingredient to competitive advantage.

7–8   These four points are noted in "The National Innovation Initiative Summit and Report," *2005 National Innovation Survey*, Council on Competitiveness, Washington, D.C., 2005, p. 38.

8   The slow shift in American R&D structures over the past 50 years is discussed in a Reed Business Information article, "Government Spending Continues to Drive R&D Growth; Fueled Mostly by Large Non-Industrial Investments, U.S. R&D Spending Is Expected to Grow Nearly 3.8% in 2005, Its Largest Forecast Improvement in Four Years," *R&D Magazine*, January 2005.

9   "Edison drove innovation on many levels . . ." Edison mastered several forms of innovation, as follows:

  • *Strategic innovation*—Launching new products using an unproven business model, often yielding entirely new industries (e.g., Edison's incandescent light bulb, development of the electrical power industry).

  • *Technological innovation*—The development of fundamentally new-to-the-world technologies (e.g., Edison's creation of the first phonograph, the first alkaline storage battery, the first motion picture camera, the first movie).

- *Product or Service innovation*—The improvement of existing technologies, or the re-combination (convergence) of technologies to offer new products or services previously unavailable (e.g., Edison's improved phonograph, the improved cylindrical record, in-home delivery of electrical power service).

- *Process innovation*—Identification of new business processes for getting things done differently and more efficiently (e.g., R&D laboratories at Menlo Park and West Orange; development of an entirely new process for mining ore).

- *Design innovation*—Development of ergonomics or aesthetics associated with product design that helps drive breakthrough consumer appeal (e.g., the glass lamp used in the light bulb, the improved phonograph, the alkaline storage battery). Source: the authors.

10   "We've got to get every member of the organization, from top to bottom, literate in innovation . . ." Personal interview, Professor Vijay Govindarajan and Sarah Miller Caldicott, November 2006. **Resource:** http://mba.tuck.dartmouth.edu/pages/faculty/vg.govindarjan/

11   "Competency" is derived from the term "core competence" coined by renowned business strategist Gary Hamel in his book *Competing for the Future*, Harvard Business School Press, Boston, MA, 1995.

12   "People who are only good with a hammer see every problem as a nail." See Maslow's classic *Toward a Psychology of Being* 2, 3rd Edition, Abraham H. Maslow, Wiley, Hoboken, NJ, 1998; and Edward Hoffman's biography *The Right to be Human*. A Biography of Abraham Maslow, McGraw-Hill, New York, 1999.

CHAPTER 2—THE STUFF OF DREAMS: THE LIFE OF THOMAS EDISON (1847–1931)

17   "Everything comes to him who hustles while he waits." Thomas A. Edison, *Edison & Ford Quote Book*, Edison & Ford Winter Estates, Fort Myers, FL, 2004, page 5.

19   "In 1889, the year in which Villiers set his novel . . ." *Edison: A Life of Invention*, Paul Israel, Wiley, Hoboken, NJ, 1998, page 364.

19   ". . . if you should tell me you could make babies by machinery, I shouldn't doubt it." Paul Israel, *Edison: A Life of Invention*, page 75.

19   "addled" brain, Paul Israel, *Edison: A Life of Invention*, page 6.

20   ". . . she was so true and sure of me . . ." Edison National Historic Site, National Park Service, Department of the Interior, www.nps.gov/edis/home_family/fam_album.htm.

20   "My mother taught me how to read good books quickly, . . ." Paul Israel, *Edison: A Life of Invention*, page 7.

20   "I can still remember the flash of enlightenment which shone from his pages." Paul Israel, *Edison: A Life of Invention*, page 8.

20   ". . . would yet blow [their] our heads off . . ." Paul Israel, *Edison: A Life of Invention*, page 11.

21   "... instead of the usual 100 papers I could sell 1000." Paul Israel, *Edison: A Life of Invention*, page 16.

21   "These early experiences as a newspaper publisher and entrepreneur set the stage for Edison's eventual success in promoting his innovations to the world." Paul Israel, *Edison: A Life of Invention*, page 17.

22   "I start where the last man left off." Thomas A. Edison, Edison & Ford Winter Estates and Museum.

23   "I have got so much to do and life is so short, I am going to hustle." Thomas Edison said this in conversation with laboratory assistant James Adams while living in Boston as a young inventor. From *Edison, His Life and Inventions*, Frank Dyer and Thomas Martin, Harper Brothers, New York, 1929, Chapter VI.

24   US Patent No. 6,469, for "Buoying Vessels over Shoals," granted in 1849. In History of the U.S. Patent Office, Kenneth W. Dobyns, 1994, Chapter 25.

25   "Edison ... followed the ideology of self-improvement to move through the ranks." Paul Israel, *Edison: A Life of Invention*, page 24.

25   He pledged that he would "never waste time inventing things that people would not want to buy." Thomas A. Edison, http://www.thomasedison.com.

26   ... he would now to "devote his time to bringing out his inventions." Paul Israel, *Edison: A Life of Invention*, page 47.

26   "People here come and buy without your soliciting." Paul Israel, *Edison: A Life of Invention*, page 48.

26   "... the best known electro-mechanician in the country." Paul Israel, *Edison: A Life of Invention*, page 49.

26   "... quite the best thing yet for taking a number of copies." Paul Israel, *Edison: A Life of Invention*, page 126.

26   Identifies the "etheric force" through experimentation that led to "sparking" between metal tubes in the laboratory. Paul Israel, *Edison: A Life of Invention*, pages 111–115, 469–470.

26   "Edison's new moonshine." Paul Israel, *Edison: A Life of Invention*, page 113.

27   "... a new model for invention that became the cornerstone of modern industrial research." Paul Israel, *Edison: A Life of Invention*, page 118.

27   "The Edison Effect" Paul Israel, *Edison: A Life of Invention*, page 469.

27   "... the best equipped and largest laboratory extant." Paul Israel, *Edison: A Life of Invention*, page 261.

28   "... does for the eye what the phonograph does for the ear." Thomas A. Edison, Edison National Historic Site, National Park Service, Department of the Interior, www.nps.gov/edis/edisonia/tae_bio.html.

28 "The Black Maria" was the world's first movie studio, designed by Edison and collaborator William Kennedy-Laurie Dickson—a photographer and experimenter in Edison's employ. The Black Maria looks like a miniature wooden frame house with a black tarpaulin drawn across sections of the upper portion of the structure. The tarp could be moved or shifted to allow sunlight into the "studio" within. The entire structure was placed on a swivel base that could be turned to provide natural sunlight as desired for filming. Dickson likely came up with the notion of the Black Maria as a huge "darkroom" that could also be used for developing the film once Edison's kinetoscope had captured images on film.

29 "I'm going to do something now so different . . ." Paul Israel, *Edison: A Life of Invention*, page 339.

29 "Well, it's all gone . . ." Dyer and Martin, *Edison, His Life and Inventions*, Chapter IX.

33 "Mr. Edison, I can always tell when . . ." Paul Israel, *Edison: A Life of Invention*, page 73.

33 ". . . shaken with grief, weeping and sobbing . . ." Paul Israel, *Edison: A Life of Invention*, page 230.

33–34 ". . . equal partnership . . ." and "lust for ownership" and "petty tasks" and "brain exercise" and "sex independence," Paul Israel, *Edison: A Life of Invention*, pages 255–256; also, "An Inventor's Wife: Mina Edison," Paul Israel, *Timeline* magazine, May-June 2001, Ohio Historical Society, pages 2–19.

34–36 For more information on the Edison/Miller family lineage, please consult the following resources: the Edison Papers Project website at http://edison.rutgers.edu, with a timeline at http://edison.rutgers.edu/famchron.htm; Pamela Miner, curator, the Edison & Ford Winter Estates, at pminer@efwefla.org; Jonathon Schmitz, archivist, Miller Family Papers at the Chautauqua Institution Archives, Chautauqua, New York, at jschmitz@ciweb.org; for history about Lewis Miller as co-founder of the Chautauqua Institution in Chautauqua, New York, see http://www.ciweb.org/history.html.

34 "He was one of the kindest and most lovable men I ever knew." Paul Israel, *Edison: A Life of Invention*, page 247.

35 ". . . gentleness and grace of manner . . ." Paul Israel, *Edison: A Life of Invention*, page 244.

35 ". . . this celestial mudball has made yet another revolution . . ." Paul Israel, *Edison: A Life of Invention*, page 246.

35 ". . . got thinking about Mina . . ." Paul Israel, "An Inventor's Wife: Mina Edison," page 4.

35 "We could use pet names . . ." Paul Israel, "An Inventor's Wife: Mina Edison," page 7.

35 "Will you marry me?" Paul Israel, *Edison: A Life of Invention*, page 247.

36 "It's a great deal too nice for me . . ." Thomas A. Edison, included in a PBS special with Edison experts Dr. Paul Israel and Neil Baldwin, www.pbs.org/wgbh/amex/edison/filmmore/transcript/index.htm.

36   "Mina Miller is the sweetest little woman . . ." Thomas A. Edison, Edison National Historic Site, National Park Service, Department of the Interior, www.nps.gov/edis/home_family/fam_album.htm.

36   "domestic engineer" Paul Israel, "An Inventor's Wife: Mina Edison," page 19.

36   ". . . tried to organize our home and our home life . . ." Paul Israel, "An Inventor's Wife: Mina Edison," pages 2–19.

36   ". . . oblivious of Sunday . . ." Paul Israel, Edison: A Life of Invention, page 246.

36   "I know this world is ruled by infinite intelligence." Thomas A Edison, Edison & Ford Winter Estates and Museum.

36   "what a wonderfully small idea mankind has of the almighty." Paul Israel, "An Inventor's Wife: Mina Edison," page 6.

37   ". . . the machine has been human being's most effective escape from bondage." Paul Israel, Edison: A Life of Invention, page 444.

37   "Nonviolence leads to the highest ethics . . ." Thomas A. Edison, set in the context of Edison's position on non-violence relative to that of other great world thinkers, web.mit.edu/justice/www/download/week1.pdf.

37   "I am proud of the fact that I never invented weapons to kill." Thomas A. Edison, Edison & Ford Quote Book, page 12.

37   "If we all try to carry out the Golden Rule . . ." Paul Israel, Edison: A Life of Invention, page 9.

37   ". . . a one man opinion on tunes is all wrong." Paul Israel, Edison: A Life of Invention, page 437.

40   ". . . his biting sarcasm . . ." Paul Israel, Edison: A Life of Invention page 276.

38–39 The media exaggerated the "rivalry" between Tesla and Edison. According to historian Dr. W. Bernard Carlson, who has studied Tesla's life and work in detail, "the real rivalry was generated by the NY newspapers in the 1890s. The papers loved to interview both Tesla and Edison and to compare their positions. Frequently, the headlines began with 'The Two Wizards.' Hence I think much of the ballyhoo about Edison vs. Tesla can be laid at the doorstep of yellow journalism." From an email sent by Dr. Carlson to Dr. Paul Israel, January 28, 1999.

40   "I have friends in overalls . . ." Thomas A. Edison, Edison & Ford Quote Book, page 14.

40   "Edison's Green Vision." Quoted in James D. Newton, uncommon Friends, Harcourt, New York, 1987, page 31.

40   "Mr. Edison is in many respects an odd man . . ." Paul Israel, Edison: A Life of Invention, page 156.

41   "A man of common sense would feel at home . . ." Paul Israel, Edison: A Life of Invention, page 156.

41   ". . . affability and playful boyishness." Paul Israel, *Edison: A Life of Invention*, page 156.

41   "His sense of humor . . ." Dyer and Martin, *Edison, His Life and Inventions*, Chapter XXIX.

CHAPTER 3—COMPETENCY #1: SOLUTION-CENTERED MINDSET

48   "I never did a day's work in all my life . . ." Thomas A. Edison, *Edison & Ford Quote Book*, page 22.

48   ". . . the money value of an invention . . ." Paul Israel, *Edison: A Life of Invention*, page 440.

48   "surprise nature into revealing her secrets . . ." Dyer and Martin, Chapter XXIV.

49   John S. Dacey and Kathleen H. Lennon, Jossey-Bass, San Francisco, CA, 1998. *Understanding Creativity: The Interplay of Biological, Psychological, and Social Factors*, Key findings are noted throughout Chapter 5: Ten Traits that Contribute to the Creative Personality. pages 98 and 111.

49   "great amounts of energy to invest intensely in their work . . ." and willingness to "persevere in the face of frustration." Dacey & Lennon, *Understanding Creativity* page 111.

49   "Edison seemed pleased when he used to run up against a serious difficulty . . ." Dyer and Martin, *Edison, His Life and Inventions*, Chapter XXII.

49–50   **Resource:** Dr. Richard Restak is a practicing neuroscientist and author. His book, *Mozart's Brain and The Fighter Pilot: Unleashing Your Brain's Potential* Richard Restak, M.D., Three Rivers Press, New York, 2002. is highly recommended as accessible reading on the brain and its diverse functions. www.RichardRestak.com

51–52   "I think that goal-setting is an important point . . ." Personal interview, Steve Odland and Sarah Miller Caldicott, August 2006.

52   "Optimistic and hopeful to a high degree . . ." Chapter XXII, Dyer & Martin.

52   "My philosophy of life: 'Work and look on the bright side of everything.'" *Thomas A. Edison, Edison & Ford Quote Book*, page 16.

53   "A lively disposition always looking on the bright side of things . . ." Paul Israel, *Edison: A Life of Invention*, page 15.

53   "I once made an experiment in Edison's laboratory . . ." Dyer and Martin, *Edison, His Life and Inventions*, Chapter XXIV.

53   "Our greatest weakness lies in giving up . . ." Thomas A. Edison, *Edison & Ford Quote Book*, page 4.

54   "I am repairing my concrete buildings . . ." Paul Israel, *Edison: A Life of Invention*, page 432.

54   Descriptions of the fire and the various estimates of Edison's material and financial losses can be found in the following documents: *New York Times*, December 10, 1914, page 1; Neil Baldwin, *Edison: Inventing the Century*, Neil Baldwin, University of Chicago, Chicago, IL, 2001, page 336; "Corporate Report, 1914," Miller Family Papers, Chautauqua Institution Archives, Chautauqua, NY.

54   "Where others might see disaster or failure . . ." and "took advantage of the latest improvements in factory design . . ." and "I am repairing my concrete buildings . . ." Paul Israel, *Edison: A Life of Invention*, page 432.

54   "Results! Why man, I have gotten a lot of results . . ." Dyer and Martin, *Edison, His Life and Inventions*, Chapter XXIV.

55   "Your notes, like your confident face . . ." Paul Israel, *Edison: A Life of Invention*, page 59.

55   "Be courageous. I have seen many depressions . . ." Thomas A. Edison, *Edison & Ford Quote Book*, page 8.

55   "Nearly every man who develops an idea . . ." Thomas A. Edison, *Edison & Ford Quote Book*, page 21.

55   "Many of life's failures are people who did not realize . . ." Thomas A. Edison, *Edison & Ford Quote Book*, page 17.

55–56   **Resource**: Dr. Martin Seligman, *Learned Optimism : How To Change Your Mind and Your Life*, Martin E. P. Seligman, Ph.D., Vintage, New York, 2006. www.positivepsychology.org.

58   Quotes from Drs. Keck, West and Langer are taken from personal interviews with Sarah Miller Caldicott, conducted in September and November 2006.

58   "His questions were so ceaseless and innumerable . . .", Dyer and Martin, *Edison, His Life and Inventions*, Chapter II.

59   "I used never to be able to get along at school . . ." *The Thomas A. Edison Album*, Lawrence Frost, Superior Publishing Co., Seattle, 1969, page 23.

59   "I was a careless boy . . ." Lawrence Frost, *The Thomas A. Edison Album*, page 23; "My mother taught me how . . ." Paul Israel, *Edison: A Life of Invention*, page 7.

59   "Certain it is that under this simple regime . . ." Dyer and Martin, *Edison, His Life and Inventions*, Chapter II.

59   "I was never able to make a fact my own without seeing it . . ." Paul Israel, *Edison: A Life of Invention*, page 96.

59   "To invent, you need a good imagination . . ." Thomas A. Edison, *Edison & Ford Quote Book*, Edison & Ford Winter Estates, page 7.

61   "After I became a telegraph operator, . . ." Lawrence Frost, *The Thomas A. Edison Album*, page 38.

61 "I didn't read a few books, I read the library." Lawrence Frost, *The Thomas A. Edison Album*, page 32.

61 "When I want to discover something, I begin by reading up . . ." Thomas A. Edison & Ford Winter Estates and Museum, museum poster, Robert Jacobsen, Puretree Media.

61–62 "Here may be found the popular magazines, . . ." Dyer and Martin, *Edison, His Life and Inventions*, Chapter XXV.

62 "The [library] shelves are . . . filled with countless thousands . . ." Dyer and Martin, *Edison, His Life and Inventions*, Chapter XXV.

62 ". . . in addition to the knowledge he has acquired . . ." Dyer and Martin, *Edison, His Life and Inventions*, Chapter XXIV.

62 "One of the main impressions left upon me . . ." Dyer and Martin, *Edison, His Life and Inventions*, Chapter XXIV.

63 "Great ideas originate in the muscles." Thomas A. Edison, *Edison & Ford Quote Book*, page 14.

64 **Resource:** ". . . the human visual system can photograph an entire page of print . . ." Tony Buzan, *The Speed Reading Book*, BBC, London, 1997. See www.buzanworld.com.

64 **Resource:** Paul Scheele is the originator of PhotoReading and a leader in the field of accelerated learning. See www.photoreading.com.

66 "I see what has been accomplished at great labor . . ." Dyer and Martin, *Edison, His Life and Inventions*, Chapter XXIV.

66 **Resource:** For further information about TRIZ and Genrich Altshuller: www.aitriz .org/ai/index.php. Also, Jack Hipple: www.innovation-triz.com.

67 "I go through the literature and say, . . ." Personal interview, Dr. Jim West and Sarah Miller Caldicott, September 2006.

67 "We don't know a millionth of a percent about anything." Thomas A. Edison, *Edison & Ford Quote Book*, page 25.

67–68 "The only way to keep ahead of the procession . . ." Dyer and Martin, *Edison, His Life and Inventions*, Chapter XXIV.

68 "Edison has proved himself a great force . . ." Paul Israel, *Edison: A Life of Invention*, page 468.

68–69 . . . an invention "every ten days, and 'a big thing' every six months . . ." Thomas A. Edison, taken from http://edison.rutgers.edu/bio-long.htm. The "invention factory" is a revolutionary concept Edison developed, which Paul Israel describes in Chapter 8 of his book, *Edison: A Life of Invention*, pages 119–141.

69 ". . . determined to have within his immediate reach . . ." Dyer and Martin, *Edison, His Life and Inventions*, Chapter XXV.

69   "the living embodiment of the spirit of the song, . . ." Dyer and Martin, *Edison, His Life and Inventions*, Chapter XXV.

71   "When asked how many experiments had been made . . ." Dyer and Martin, *Edison, His Life and Inventions*, Chapter XXII.

71   "When I have fully decided that a result is worth getting, . . ." Thomas A Edison, *Edison & Ford Quote Book*, page 15.

71   "surprise Nature into a betrayal of her secrets . . ." Dyer and Martin, *Edison, His Life and Inventions*, Chapter XXIV.

71   "ransack the jungles of the Far East . . ." Dyer and Martin, *Edison, His Life and Inventions*, Chapter XIII.

71   "Your trip to China and Japan on my account . . ." Dyer and Martin, *Edison, His Life and Inventions*, Chapter XIII.

71–72   "It is doubtful whether, in the annals of scientific research and experiment, . . ." Dyer and Martin, *Edison, His Life and Inventions*, Chapter XIII.

72   "There's something wrong with this . . ." Dyer and Martin, *Edison, His Life and Inventions*, Chapter XXIV.

72   ". . . changed into a stringy, cohesive, . . ." Dyer and Martin, *Edison, His Life and Inventions*, Chapter XXIV.

72   "There is always a way to do it better . . . find it." Thomas A. Edison, Edison & Ford Winter Estates and Museum.

72   "Genius is one percent inspiration . . ." Dyer and Martin, *Edison, His Life and Inventions*, Chapter XXIV.

73   "A popular idea of Edison that dies hard is . . ." Dyer and Martin, *Edison, His Life and Inventions*, Chapter XI.

75   "Iterate, iterate, iterate." Curtis Carlson and William Wilmot, in personal conversations with the authors, and from *Innovation: The Five Disciplines for Creating What Customers Want*, Curtis R. Carlson and William W. Wilmot, Crown Business, New York, 2006.

77   "in their first go-round." From "Creativity Overflowing," by Michael Arndt, *Business Week*, May 8, 2006.

78   "They noticed a spark passing between the cores of the magnet . . ." Paul Israel, *Edison: A Life of Invention*, page 111.

78   ". . . had always attributed it to induction . . ." Paul Israel, *Edison: A Life of Invention*, page 111.

78   ". . . soon found they could get the spark by touching . . ." Paul Israel, *Edison: A Life of Invention*, page 111.

78 "... from pipes anywhere in the room," and "by placing a piece of metal ..." Paul Israel, *Edison: A Life of Invention*, page 111.

78 "... the cause of the spark is a true unknown force." Paul Israel, *Edison: A Life of Invention*, page 111.

79–80 "Our thinking tends to be hazy ..." Quoted in *The Thinker Way*, John Chafee, Little, Brown, Baston, 1998, page 17.

80–81 **Resource:** *Six Thinking Hats* Edward de Bono, 2nd Edition, Penguin, New York, 2000. www.edwdebono.com.

81 "For me, because I'm an experimentalist, ..." Personal interview, Dr. Jim West and Sarah Miller Caldicott, September 2006.

CHAPTER 4—COMPETENCY 2: KALEIDOSCOPIC THINKING

83 "[Edison has a] remarkable kaleidoscopic brain ..." Paul Israel, *Edison: A Life of Invention*, page 67.

83 "I would love to live about 300 years. ..." Thomas A. Edison, *Edison & Ford Quote Book*, page 28.

83 "Edison's inexhaustible resources and fertility of imagination ..." Dyer and Martin, *Edison, His Life and Inventions*, Chapter XXIV.

83 "... mental kaleidoscope" and "obtain a new combination of ideas ..." Lawrence Frost, *The Thomas A. Edison Album*, page 87.

84 "... in the habit of using small pocket notebooks ..." Paul Israel, *Edison: A Life of Invention*, pages 36–37.

84–85 "... such a record would be essential ..." and "all new inventions, I will here after keep a full record." Paul Israel, *Edison: A Life of Invention*, page 56.

86 "... a series of four notebooks ..." and "to be used in any contest or disputes ..." and "any ideas contained in this book ..." Paul Israel, *Edison: A Life of Invention*, page 67.

86 "I do not wish to confine myself to any particular device." Paul Israel, *Edison: A Life of Invention*, page 67.

86 "I have struck a big bonanza." *Edison's Electric Light: Biography of an Invention*, Robert Friedel and Paul Israel, Rutgers University Press, New Brunswick, NJ, 1987, page 8.

88 "... highly condensed language of thought ...", Vera John-Steiner, *Notebooks of the Mind: Explorations of Thinking*, Oxford University Press, New York, 1997, page 111.

88 "... a single word is so saturated with sense ..." Vera John-Steiner, *Notebooks of the Mind: Explorations of Thinking*, page 113.

89   "... makes it possible to gallop ahead, ..." Vera John-Steiner, *Notebooks of the Mind: Explorations of Thinking*, page 112.

89–90   **Resource:** For more guidance on using a notebook: The Intensive Journal Method created by Dr. Ira Progoff. This is a profound way to use the journaling process to develop the intuitive knowledge that is the wellspring of kaleidoscopic thinking. See www.intensivejournal.org. And *How to Think Like Leonardo DaVinci Workbook*, Michael J. Gelb, Dell, New York, 1999.

90–91   "Writing down my ideas frees up my brain from remembering so many details ..." and "For two years, our team was focusing on ..." Personal interview with Dr. John Wai and Sarah Miller Caldicott, December 2006.

91   "To have a great idea, have a lot of them." Dyer and Martin, *Edison, His Life and Inventions*, Chapter XXIV.

92   "... produced page after page of possible approaches ..." Paul Israel, *Edison: A Life of Invention*, page 134.

92   "Edison can think of more ways of doing a thing ..." Dyer and Martin, *Edison, His Life and Inventions*, Chapter XXIV.

92   "Mr. Edison turned to him quickly and said: ..." Dyer and Martin, *Edison, His Life and Inventions*, Chapter XXIV.

92   "I speak without exaggeration when I say ..." George Parsons Lathrop, "Talks with Edison," *Harper's New Monthly Magazine*, Volume 80, Issue 477, February 1890, page 434. If you desire to see the entire 10-page text, conduct a search on the landing page of the Making of America project sponsored by Cornell University, at http://cdl .library.cornell.edu/moa/moa_adv.html. Set up your search to include "Journals only," and retrieve "Talks with Edison" as the title, between 1890 and 1895.

93   "Ah, Shakespeare. That's where you get the ideas!" Paul Israel, *Edison: A Life of Invention*, page 29.

93   "'a logical mind that sees analogies' ..." Paul Israel, *Edison: A Life of Invention*, page 68.

93   "treated the magnetic lines of force ..." Paul Israel, *Edison: A Life of Invention*, page 176.

93   "... as a form of telegraph." Paul Israel, *Edison: A Life of Invention*, page 144.

93   "... envisioned using a recorder ..." Paul Israel, *Edison: A Life of Invention*, page 144.

93   The full text of this quote, taken from the "American Memory" section of the Library of Congress website, is "I am experimenting upon an instrument which does for the eye what the phonograph does for the ear, which is the recording and reproduction of things in motion ..." http://memory.loc.gov/ammem/edhtml/edmvhm.html. More information about Edison's contributions to motion pictures can be found on the Rutgers University website at this link: http://edison.rutgers.edu/pictures.htm, as well as at www.kino.com/ edison/hp.html#.

93 "Contemporary research into the nature of practical intelligence confirms that analogy is one of the mind's most powerful problem-solving tools. See Vera John-Steiner, *Notebooks of the Mind: Explorations of Thinking*, page 191.

93 "to generate analogies both within and across disciplinary boundaries." Vera John-Steiner, *Notebooks of the Mind: Explorations of Thinking*, page 192.

94 "But in addition to being extremely practical in his thoughts . . ." George Parsons Lathrop, "Talks with Edison," *Harper's New Monthly Magazine*, page 434.

94 To view the surviving 33 pages from Edison's notes for the novel *Progress*, go to: http://edison.rutgers.edu/NamesSearch/SingleDoc.php3?DocID = D9004AFW.

95 "If you want your brain to function optimally, . . ." Richard Restak, *Mozart's Brain and the Fighter Pilot*, pages 33–34.

95 ". . . imagination is based on human capabilities . . ." Bill Welter and Jean Egmon, *The Prepared Mind of a Leader: Eight Skills Leaders Use to Innovate, Make Decisions, and Solve Problems*, Jossey-Bass, San Francisco, CA, 2006, page 111.

97 "Just like soap, the polymer we developed . . ." Personal interview, Dr. Robert Langer and Sarah Miller Caldicott, September 2006.

97 "One day I was just watching this TV show on PBS. . . ." Personal interview, Dr. Robert Langer and Sarah Miller Caldicott, September 2006.

97–99 **Resource:** For more information on Image Streaming contact Dr. Win Wenger, www.winwenger.com/imstream.htm.

99 "write down what was coming and imagine what wasn't coming." Paul Israel, *Edison: A Life of Invention*, page 34.

100 "It is astonishing what an effort it seems to be . . ." Thomas A. Edison, *Edison & Ford Quote Book*, page 25.

100 "Edison would examine the tabulated test sheets, . . ." Dyer and Martin, *Edison, His Life and Inventions*, Chapter XXIV.

101 Please see the description of how Edison conceived of the telegraph stylus as a means to transfer sound waves onto a hard surface. Lawrence Frost, *The Thomas A. Edison Album*, pages 66–67.

102–104 **Resource:** For the definitive work on Mind Mapping® : *The Mind Map Book* by Tony Buzan: www.buzanworld.com.

103 ". . . integrate a large amount of apparently unrelated data . . ." and ". . . as I began to make connections between . . ." Letter to Michael J. Gelb, March 1992.

104 "The ability to discern patterns . . ." Letter from Professor James Clawson to Michael J. Gelb, March 2007.

104 "The process of pattern-seeking is very visual . . ." Personal interview, Dr. John Wai and Sarah Miller Caldicott, December 2006.

107   "... the scientist needs them in order to ..." Vera John-Steiner, *Notebooks of the Mind: Explorations of Thinking*, page 88.

109–110   **Resource:** For more information on how you can learn to draw *well enough*: *The Natural Way to Draw* by Kimon Nicolaides, *Drawing on the Right Side of The Brain* by Dr. Betty Edwards: www.drawright.com. *Conni Gordon a*uthor of more than 16 art instruction books: www.thinklearncreateandlive.com And, "The Beginner's da Vinci Drawing Course," Michael J. Gelb, *How to Think Like Leonardo DaVinci* pages 263–305.

110   **Resource:** For more information on Graphic Facilitators: Grove Consultants International: www. grove.com and The Association for Visual Practitioners: *www.visualpractitioner.org*.

110   "I'm very dyslexic, so I can't read something ..." Personal interview, Dr. Jim West and Sarah Miller Caldicott, September 2006.

112   "I shall make the electric light so cheap ..." Thomas A. Edison, *Edison & Ford Quote Book*, page 3.

112   "... an absolute ignis fatuus." Dyer and Martin, *Edison, His Life and Inventions*, Chapter XI.

112   "Much nonsense has been talked in relation to this subject ..." Dyer and Martin, *Edison, His Life and Inventions*, Chapter XI.

112   ... practical incandescence was "... utterly impossible." Dyer and Martin, *Edison, His Life and Inventions*, Chapter XXVIII.

113   "requires developing the guts and the courage ..." Bill Welter and Jean Egmon, *The Prepared Mind of a Leader*, page 126.

113   "There is no expedient to which a man will not go to avoid the labor of thinking." Thomas A. Edison, *Edison & Ford Quote Book*, page 26.

113   "There must be a champion who ..." Curt Carlson and Bill Wilmot, *Innovation: The Five Disciplines for Creating What Customers Want*, page 157.

113   "No champion, no project, no exception." Curt Carlson and Bill Wilmot, *Innovation: The Five Disciplines for Creating What Customers Want*, page 157.

113–114   "One of the things that you find out early on ..." and "Dr. Bob Mauer hired me ..." and "We later found out that Bell Labs ..." Personal interview, Dr. Donald Keck and Sarah Miller Caldicott, November 2006.

CHAPTER 5—COMPETENCY #3: FULL-SPECTRUM ENGAGEMENT

115   "He has in a degree which is literally startling ..." George Parsons Lathrop, "Talks with Edison," *Harper's New Monthly Magazine*, page 425.

116   This definition of "flow" taken from *Creativity: Flow and the Psychology of Discovery and Invention*, Mihaly Csikszentmihalyi, Harper Collins, New York, 1996, pages 112–113.

116   "When experimenting at Menlo Park . . ." Dyer and Martin, *Edison, His Life and Inventions*, Chapter XXV.

116–117   "Practical jokes, tests of strength, . . ." Paul Israel, *Edison: A Life of Invention*, pages 193–194.

117   "Our [midnight] lunch always ended with a cigar . . ." Dyer and Martin, *Edison, His Life and Inventions*, Chapter XII.

117   Edison enjoyed "coffee, pie . . ." Friedel and Israel, *Edison's Electric Light: Biography of an Invention*, page 147.

118   "As one is about to pass out of the library . . ." Dyer and Martin, *Edison, His Life and Inventions*, Chapter XXV.

118   "Sometimes when Mr. Edison had been working long hours . . ." Dyer and Martin, *Edison, His Life and Inventions*, Chapter XXII.

118   "In trying to perfect a thing, . . ." Lawrence Frost, *The Thomas A. Edison Album*, page 106.

118   Using one activity "as a relief from another . . ." George Parsons Lathrop, "Talks with Edison," *Harper's New Monthly Magazine*, page 426.

118   Edison used the term "loafing" to describe time on vacation, or idle time away from the office. Dyer and Martin, *Edison, His Life and Inventions*, Chapter XXV.

118–119   Edison was known to sit for hours fishing with "a baitless hook." Lawrence Frost, *The Thomas A. Edison Album*, page 132.

119   The *Reminiscence Effect* in *The Brain Book* by Peter Russell.

119   "Our most fundamental need as human beings . . ." Jim Loehr and Tony Schwartz, *The Power of Full Engagement*, The Free Press, NY, 2003, page 86.

119–121   **Resource:** Jim Loehr, c/o Human Performance Institute, 9757 Lake Nona Road, Orlando, Florida 32827. You can contact Jim online at info@energyforperformance.com.

121   "Edison is always himself . . ." George Parsons Lathrop, "Talks with Edison," *Harper's New Monthly Magazine*, page 425.

121   "Never is he so preoccupied or fretted with cares . . ." Dyer and Martin, *Edison, His Life and Inventions*, Chapter XXIX.

121   "He is capable of great jollity . . ." George Parsons Lathrop, "Talks with Edison," *Harper's New Monthly Magazine*, page 425.

122   "Celebrities of all kind and distinguished foreigners— . . ." Dyer and Martin, *Edison, His Life and Inventions*, Chapter XXV.

122   "I was very fond of stories and had a choice lot . . ." Robert Friedel and Paul Israel, *Edison's Electric Light: Biography of an Invention*, page 147.

122   "wishing we could tell everything as he can . . ." Paul Israel, *Edison: A Life of Invention*, page 247.

122   "One evening Edison, while spitting vigorously . . ." Lawrence Frost, *The Thomas A. Edison Album*, page 93.

123   . . . he "never blackened his boots and seldom combed his hair . . ." Lawrence Frost, *The Thomas A. Edison Album*, page 28.

123   "discolored from some chemical" Lawrence Frost, *The Thomas A. Edison Album*, page 91.

123   "all his dirt and grease on my nice white counter panes . . ." Paul Israel, *Edison: A Life of Invention*, page 124.

123   Individuals appearing in this photo of Lewis Miller's summer home at 24 Miller Park in Chautauqua, New York are, from left to right (on the porch): Thomas Edison's eldest daughter Marion Edison, Ira Miller (older brother of Mina Miller), Mrs. Gilliland, and Mary Valinda Miller. Standing near the front door, at left, is George Vincent, with Lewis Miller at right of the door (Mina's father) and business associate Ezra Gilliland seated in front at left, at the top of the stairs. Seated on the steps at left is Theodore Miller (Mina's youngest brother) with sister Grace Miller seated at right, John Vincent Miller standing near his bicycle at right below the steps, and seated in a rocking chair on the porch at right, Thomas A. Edison. At left near the tent is a woman whose name is not known, with Edison's colleague Charles Batchelor seated on the porch rail next to Mina Miller, with Mina wearing a white dress and bonnet; Edison's colleague Samuel Insull sits at the far right. The individuals on the upper porch balcony are unidentified. Miller Family archives indicate identifications were made by Madeleine Edison Sloane and Lewis Miller II, decades after the photo was taken.

123   "My job has always been to take care of Mr. Edison . . ." Paul Israel, "An Inventor's Wife: Mina Edison," *Timeline*, page 19.

123   "Maturity is often more absurd than youth . . ." Thomas A. Edison, http://www.quotableedison.com/allquotes.php.

124   "An increasing number of case studies support . . ." Letter from Professor James Clawson to Michael J. Gelb, March 2007.

124   "His laugh, in fact, is sometimes almost aboriginal; . . ." Dyer and Martin, *Edison, His Life and Inventions*, Chapter XXIX.

124–125   **Resource:** Dr. Madan Kataria, founder of Laughter Clubs Movement, author of *Laugh for No Reason*. at www.laughteryoga.org. Also, Inventables, a Chicago-based company that inspires new ways of thinking about standard, existing objects—including your own products. Contact founder Zach Kaplan at www.inventables.com.

125   "My philosophy of life is work—bringing out the secrets of nature. . . ." Thomas A. Edison, *Edison & Ford Quote Book*, page 5.

125   "to secure if possible the science of the thing." Paul Israel, *Edison: A Life of Invention*, page 209.

125   "continued innovation" as "the best means of defeating the competition." Edison believed that Paul Israel, *Edison: A Life of Invention*, page 209.

125   "At the core of his strategy . . ." Paul Israel, *Edison: A Life of Invention*, page 209.

126   ". . . would require me to give my personal attention . . ." Paul Israel, *Edison: A Life of Invention*, page 209.

126   "patent law favored infringers . . ." Paul Israel, *Edison: A Life of Invention*, page 318.

126   "After a thing is perfected . . ." Paul Israel, *Edison: A Life of Invention*, page 318.

126   "There is bound to be a delay; . . ." Paul Israel, *Edison: A Life of Invention*, page 318.

127   The table containing Dyer and Martin's estimates of the investment value of Edison's U.S. patents is in Chapter XXVII of their book, *Edison, His Life and Inventions*. The table is entitled, "Statistical Resume (approximate) of Some of the Industries in the United States Directly Founded Upon or affected by Inventions of Thomas A. Edison." The storage battery is not included in Dyer and Martin's calculations because the improved battery had just been released in 1910, and thus no industry records had yet been established for it. The storage battery became one of Edison's most profitable innovations. Figures were brought into current dollars for this calculation and all others noted in the book using the GDP deflator at the following link: http://measuringworth.com/calculators/uscompare/

127   ". . . become extremely skeptical as to the value of . . ." Paul Israel, *Edison: A Life of Invention*, page 318.

127   ". . . particularly objected to the fact that . . ." and "cost him his basic patents on the phonograph . . ." Paul Israel, *Edison: A Life of Invention*, page 318.

128   Edison's work on the armature was revolutionary, and caused a major stir in the engineering and scientific communities. This is recapped by Paul Israel, *Edison: A Life of Invention*, page 415.

128   ". . . increased [the] capacity of platinum to withstand . . ." Paul Israel, *Edison: A Life of Invention*, page 183.

128   "all the leading universities and colleges . . ." Paul Israel, *Edison: A Life of Invention*, page 464.

129   "These men all had complete faith in his ability . . ." Dyer and Martin, *Edison, His Life and Inventions*, Chapter XXII.

129   "The world is full of people who want to . . ." Personal interview, Dr. Vijay Govindarajan and Sarah Miller Caldicott, November 2006.

130   Dr. Henry Chesbrough is widely regarded as the individual who coined the term "Open Innovation." Dr. Chesbrough authored a book on the subject, entitled: *The New Imperative for Creating and Profiting from Technology*, Harvard Business School Press, Cambridge, MA, 2003.

130   "The Future of Outsourcing: How It's Transforming Whole Industries and Changing the Way We Work," by Pete Engardio, with Michael Arndt and Dean Foust, *Business Week*, January 30, 2006.

130   "We've been pushing hard for improving patent quality . . ." Personal interview, Michael Wing and Sarah Miller Caldicott, November 2006.

131–132   **Resource:** The U.S. Patent and Trademark Office website at www.uspto.gov. For a more global perspective on protecting your Intellectual Property, the World Intellectual Property Organization (WIPO) is a specialized agency of the United Nations dedicated to developing a balanced and accessible international intellectual property system. WIPO's mission is to facilitate the evolution of a global IP system that will "reward creativity, stimulate innovation and contribute to economic development while safeguarding the public interest." You can learn about the status of patents internationally through their website: www.wipo.int/portal/index.html.en.

132   ". . . especially in the form of PCT patent applications . . ." Personal interview, Dr. John Wai and Sarah Miller Caldicott, December 2006.

133   "clear cut and direct" and "minutest exactitude" Dyer and Martin, *Edison, His Life and Inventions*, Chapter XXIV.

134   "When the plant was nearly ready . . ." Lawrence Frost, *The Thomas A. Edison Album*, page 112.

134   "When I was presented to Mr. Edison, . . ." Dyer and Martin, *Edison, His Life and Inventions*, Chapter XIII.

135   "instant grasp" Dyer and Martin, *Edison, His Life and Inventions*, Chapter XXIV.

135   "Anything less than a conscious commitment . . ." *First Things First*, Stephen Covey, Simon & Schuster, New York, 1994, page 32.

136   **Resource:** Jon Kabat-Zinn's books and audio programs, and courses offered by the Center for Mindfulness in Medicine, Health Care, and Society: *Wherever You Go, There You Are: Mindfulness Meditation in Everyday Life; Mindfulness for Beginners* (Audio CD) by Jon Kabat-Zinn. The Center for Mindfulness in Medicine, Health Care, and Society: www.umassmed.edu/cfm/index.aspx.

136   "The best managers are indeed jugglers . . ." article by Professor Arne May, *Nature*, January 22, 2004, pages 427, 311–312.

136   **Resource:** The International Juggler's Association: www.juggle.org.

136   "I work on ten or more game projects at a time . . ." Personal interview, Joel Jaffe and Sarah Miller Caldicott, December 2006.

137   "The best thinking has been done in solitude. . . ." Thomas A. Edison, *Edison & Ford Quote Book*, page 18.

137   "the ease and rapidity with which [Edison] adjusts himself . . ." and ". . . due to ready and absolute control of his mental forces." George Parsons Lathrop, "Talks with Edison," *Harper's New Monthly Magazine*, page 425.

138   "No. 12 [is] Edison's favorite room, . . ." Dyer and Martin, *Edison, His Life and Inventions*, Chapter XXV.

138   "Edison himself flits about . . ." Paul Israel, *Edison: A Life of Invention*, page 191.

140   "My research implies . . ." Dr. Peter Suedfeld, www.psych.ubc.ca/~psuedfeld.

140–141   **Resource:** *Solitude: A Return to the Self*, Anthony Storr; Free press, New York, 2005; *Celebrating Time Alone: Stories of Splendid Solitude*, Lionel Fisher, Atria Books, New York, 2001; *The Call of Solitude* Ester Buchholz, Ph.D., Simon: Schuster, New York, 1999. Also, learn more about the REST center locations in the United States at www.float dreams.com/wheretofloat.htm.

CHAPTER 6—COMPETENCY #4: MASTER-MIND COLLABORATION

144   ". . . at the age of 21, [John] asked for a job . . ." Lawrence Frost, *The Thomas A. Edison Album*, page 55.

144   Paul Israel explains what Edison was seeking in his new hires: "he preferred men whom he considered generalists . . . he wanted to train them himself, and evidently he wanted to pick men . . . that were observant and generally interested in things.'" pgs. 53–54, "Inventing Industrial Research: Thomas Edison and the Menlo Park Laboratory," Paul B. Israel, *Endeavor*, Volume 26(2), 2002, pages 53–54.

144–145   ". . . an electrician by training." Paul Israel, *Edison: A Life of Invention*, page 275.

145   "taught me the right way to experiment." Paul Israel, *Edison: A Life of Invention*, page 275.

145   The full text of the letter from Grosvenor P. Lowrey to Edison can be found at: http://edison.rutgers.edu/NamesSearch/SingleDoc.php3?DocId = D7913L.

146   Samples of Edison's "Mental Fitness Tests" can be found at: www.nps.gov/archive/edis/edifun/quiz/quizhome.htm

146   "Men who have gone to college I find amazingly ignorant. . . ." Thomas A. Edison, within the presentation of Edison's Mental Fitness Test, http://www.nps.gov/archive/edis/edifun/quiz/qu_intro.swf

146   It was "not the money they want, but a chance for their ambition to work." Paul Israel, *Edison: A Life of Invention*, page 324.

148   "We work hard to align the interviewee's perception of us . . ." Personal interview, Dr. Richard Sheridan and Michael J. Gelb, November 2006.

148   "Get the right people on the bus, . . ." Jim Collins, *Good to Great: Why Some Companies Make the Leap . . . and Others Don't*, Harper Business, New York, 2001, page 191.

148   Lewis Latimer was an African-American draftsman, inventor, and engineer hired by Edison, who worked in the legal department of the Edison Electric Light Company as chief draftsman and patent expert. In this capacity, he drafted drawings and documents related to Edison patents, inspected plants in search of infringers of Edison's patents, conducted patent searches, and testified in court proceeding on Edison's behalf. He also wrote one of the world's most thorough books on electric lighting, *Incandescent Electric Lighting: A Practical Description of the Edison System* (out of print). Lewis was named one of the charter members of the Edison Pioneers, a distinguished group of people deemed responsible for creating the electrical industry. For more information on Lewis Latimer, go to: http://www.blackinventor.com/pages/lewislatimer.html.

149   "the sole directing mind" Paul Israel, *Edison: A Life of Invention*, page 195.

150   "In this respect of collaboration, . . ." Dyer and Martin, *Edison, His Life and Inventions*, Chapter XIV.

151   "weaving together of ideas, styles of work, . . ." and ". . . for each other's blind spots." Vera John-Steiner, *Notebooks of the Mind: Explorations of Thinking*, page 187.

151   "Collaboration operates through a process in which . . ." and "visualizing a solution by working with ideas . . ." Vera John-Steiner, *Notebooks of the Mind: Explorations of Thinking*, pages 187–188.

152   **Resource:** "enabling organizations to transform their cultures . . ." "Multidisciplinary teams are the heart . . ." www.ideo.com Please also see *The Ten Faces of Innovation*, by Tom Kelley.

152   **Resource:** *The Enneagram*, four CDs by Dr. Dennis Perman: www.shop.themaster-scircle.com/theenneagram.html. To discover your type we recommend: The *Riso-Hudson Enneagram Type Indicator* www.enneagraminstitute.com. For further information on brain dominance, we recommend Hermann International at www.hbdi.com.

153   "As technology gets more complex . . ." Personal interview, Dr. Jim West and Sarah Miller Caldicott, September 2006; and "My lab has people with 10 to 12 different disciplines . . ." Personal interview, Dr. Robert Langer and Sarah Miller Caldicott, September 2006.

153   "where people continually expanding their capacity . . ." *The Fifth Discipline: The Art & Practice of the Learning Organization*, Peter Senge, Doubleday, New York 1990, page 3.

154   "The laboratory initiated the work of the greater organization . . ." Paul Israel, *Edison: A Life of Invention*, page 399.

154   In this photo of the second floor of Edison's Menlo Park Laboratory, appearing from left to right, are: Ludwig K. Boehm, Charles L. Clarke, Charles Batchelor, William Carman, Samuel D. Mott, George Dean, Thomas A. Edison (wearing artisan's cap and kerchief), Charles T. Hughes, George Hill, George E. Carman, Francis Jehl, John W. Lawson, Charles Flammer, Charles P. Mott, and James MacKenzie, who taught Edison telegraphy. The pipe organ shown in the background was a gift to Edison from investor Hilborne Roosevelt.

155   "Those who were gathered around him . . ." Dyer and Martin, *Edison, His Life and Inventions*, Chapter XII.

155   . . . the "fluency of ideas and flexibility of approach . . ." Vera John-Steiner, *Notebooks of the Mind: Explorations of Thinking*, page 187.

155   This segment is adapted from Welter and Egmon's *The Prepared Mind of a Leader*, page 126.

156   "I generally instructed them on . . ." Paul Israel, *Edison: A Life of Invention*, page 192.

156   These passages include: text from a combination of sources, primarily a personal interview with Michael Wing and Sarah Miller Caldicott in December 2006; data taken from an interview with Sam Palmisano entitled, "Leading Change When Business Is Good," Paul Hemp and Thomas A. Stewart; *Harvard Business Review*, December 2004; data from "IBM Goes Public to Collect Ideas," Jon Van, *Chicago Tribune*, October 1, 2006; and, data taken from the 2006 IBM Annual Report supplement, the "Think Book," at www.ibm.com/annualreport/.

157   "by focusing the attention of an organization . . ." Dr Diana Whitney, conversation with Michael J. Gelb, October 2006.

157–158   **Resource:** For further information on Appreciative Inquiry, we recommend *The Power of Appreciative Inquiry: A Practical Guide to Positive Change*, Diana Whitney, Amanda Trosten-Bloom, and David Cooperrider, Berrett-Kochler, *San Francisco*, 2003. www.positivechange.com.

158   "Edison was a master at addressing resistance . . ." Personal interview, Steve Odland and Sarah Miller Caldicott, August 2006.

158   "The best reward is not to give money, but to . . ." Dr. Clotaire Rapaille, *Seven Secrets of Marketing*, Executive Excellence Publishing, Utah, 2001, page 26.

159   "Edison made your work interesting . . ." www.americanhistory.si.edu/lighting.

159   "had been 'indoctrinated' in his methods . . ." and "a very considerable opportunity . . ." and "after working under close instructions . . ." and "got to understand his methods pretty well." Paul Israel, *Edison: A Life of Invention*, page 275.

159   Paul Israel describes how Edison would lecture weekly on a variety of topics, delivering his talks from the foyer of the West Orange laboratory. Paul Israel, *Edison: A Life of Invention*, page 274.

160   "one of the finest technical and scientific libraries in the world." Paul Israel, *Edison: A Life of Invention*, page 274.

160   "used to study mathematics together . . ." Paul Israel, *Edison: A Life of Invention*, page 274.

160   "The laboratory's machinists could . . . add to their earnings . . ." Paul Israel, *Edison: A Life of Invention*, page 274.

160–161   Paul Israel discusses the royalty arrangements between Edison and various members of his staff. Paul Israel, *Edison: A Life of Invention*, page 195.

160–161   It was important in Edison's work to identify who had the lead role in the development of an idea. This became crucial not only for reasons of royalty sharing, but for legal recognition in the patent itself. Edison was generally recognized as "the sole directing mind." Although in some instances he could likely have named more co-inventors, "joint patents were open to challenge under the existing patent law." Paul Israel, *Edison: A Life of Invention*, page 195.

161   "know that if I am successful that I don't keep it all for myself." Paul Israel, *Edison: A Life of Invention*, page 274.

162   "Rewards come in many forms. . . ." Carlson and Wilmot, *Innovation: Five Disciplines for Creating What Customers Want*, page 193.

163   "When I joined the pharmaceutical industry, . . ." Personal interview, Dr. Annaliesa Anderson and Sarah Miller Caldicott, December 2006.

164   "We're on a journey toward fully rewarding collaboration . . ." Personal interview, Mike Wing and Sarah Miller Caldicott, December 2006.

166   ". . . kept posted, and knew from their activity . . ." Dyer and Martin, *Edison, His Life and Inventions*, Chapter V.

167   "There they amazed the staff by playing the little machine . . ." Paul Israel, *Edison: A Life of Invention*, page 145.

167   "The Wizard of Menlo Park" Paul Israel, *Edison: A Life of Invention*, page 147.

168   "exquisite ingenuity and . . . usefulness." Paul Israel, *Edison: A Life of Invention*, page 125.

169   "every day some 30,000 people heard some twenty-five phonographs . . ." Paul Israel, *Edison: A Life of Invention*, page 371.

169   "a young man who has made a gas car." Lawrence Frost, *The Thomas A. Edison Album*, page 126.

169   "You have it! Keep at it!" Lawrence Frost, *The Thomas A. Edison Album*, pages 126–127.

169–170   "The Chinese call it 'guanxi.' And they believe it to be . . ." Personal interview with Dr. Jim Clawson and Michael J. Gelb, March 2007. **Resources:** JimClawson@virginia.edu/

170   "My admonition to my scientists was, 'Get out around the world . . .'" Personal interview with Dr. Donald Keck and Sarah Miller Caldicott, November 2006.

170   **Resource:** *The Secrets of Savvy Networking*, Susan RoAne, Grand Central Publishing, *New York*, 1993. www.susanroane.com.

171   "Organizational theorists have known for decades . . ." Personal interview, Professor Jim Clawson and Michael J. Gelb, March 2007.

171   **Resource:** You can learn more about an online social networking tool developed by Professor Rob Cross of the Darden School of Business at the University of Virginia by going to: https://webapp.comm.virginia.edu/networkroundtable/.

CHAPTER 7—COMPETENCY #5: SUPER-VALUE CREATION

173   "the process of creating and delivering new customer value . . ." Carlson and Will Wilmot, *Innovation: The Five Disciplines for Creating What Customers Want, page 6.*

177   Items mentioned taken from "Inventions of the Year," *Time,* November 13, 2006.

177–180   **Resource:** *The Long Tail,* Chris Anderson, Hyperion, New York, 2006; *The Age Wave,* Ken Dychtwald, Ph.D., and Joe Flower, Bantam, New York, 1990; *Small Is the New Big,* Seth Godin, Portfolio, New York, 2006; *The World in 2020,* Hamish McRae, Harrard Business School, Cambridge, MA, 1996; *Mind Set!* John Naisbitt, Harper Collins, New York, 2006; *Revolutionary Wealth,* Alvin Toffler and Heidi Toffler, Knopf, New York, 2006; *Dictionary of the Future* Faith Popcorn and Adam Hanft, Hyperion, New York, 2001.

178   Some "Increasing Speed" trend data is adapted from information contained in the "Smart Objects" segment of the IBM 2005 Annual Report supplement, page 14. The entire supplement can be viewed at www.ibm.com/annualreport/; click on "It's A Great Time To Be An Innovator."

178   "Smarter Objects" segment includes data adapted from information contained in the IBM 2005 Annual Report supplement, "Smart Objects," page 14. The entire supplement can be viewed at www.ibm.com/annualreport/ by clicking on "It's A Great Time To Be An Innovator."

179   "More Collaboration and Co-creation" includes data adapted from information contained in the 2005 IBM Annual Report supplement, page 16. The entire supplement can be viewed at www.ibm.com/annualreport/, and click on "It's A Great Time To Be An Innovator."

180   **Resource:** To learn more about how you can stay on top of trends using online resources, consult: www.blogpulse.com; and www.trendsight.com

184   "When I first met Heidi Klum in 1996, . . ." Letter from Desiree Gruber to Sarah Miller Caldicott, December 2006.

185   "Anything that won't sell, I don't want to invent. . . ." Thomas A Edison, *Edison & Ford Quote Book*, page 7.

185–186   "Thomas Edison was a master at knowing when to attack a problem . . ." Carlson and Wilmot, *Innovation: The Five Disciplines for Creating What Customers Want*, page 61.

186   "He had high hopes of finding a ready market . . ." Paul Israel, *Edison: A Life of Invention*, page 107.

188   "carry the outfit from office to office . . ." and "mechanical defects in the pen . . ." Paul Israel, *Edison: A Life of Invention*, page 107.

188   "identified fifteen mechanical refinements . . ." Paul Israel, *Edison: A Life of Invention*, page 107.

188   "Indeed, it has been difficult to supply the demand." Paul Israel, *Edison: A Life of Invention*, page 108.

188–189   Paul Israel details Edison's process for canvassing and evaluating communities desiring a new power plant, *Edison: A Life of Invention*, pages 224–225.

189   "The serious and important part of the mail, . . ." Dyer and Martin, *Edison, His Life and Inventions*, Chapter XXV.

191   "For over twenty years, I have been working . . ." Personal interview, Dr. Rich Sheridan and Michael J. Gelb, November 2006.

191–192   **Resource:** To learn more about how ethnographic research connects to innovation, consult Chicago-based product design experts Gravity Tank at www.gravitytank.com.

197–198   Strategic innovation and business model experts Vijay Govindarajan and Chris Trimble of the Amos Tuck School of Business, in their book *Ten Rules for Strategic Innovators: From Idea to Execution*, identify ten powerful rules for large companies to follow when undertaking innovation efforts. This segment primarily focuses on concepts discussed in Chapters 1 through 4 of their book. *Ten Rules* was released in 2005 by the Harvard Business School Press, and is highly recommended reading.

199–200   "I have been figuring during the past week on some estimates . . ." Friedel and Israel, *Edison's Electric Light: Biography of an Invention*, page 122.

200   Edison made extensive calculations to forecast the trajectory of the lighting industry. For a further discussion of this, see Friedel and Paul Israel, *Edison's Electric Light: Biography of an Invention*, Chapters 6 and 7.

200   "To develop your business model you have to do your homework." Carlson and Wilmot, *Innovation: The Five Disciplines for Creating What Customers Want*, page 150.

200   "Until . . . there is a viable business solution there is no opportunity." Carlson and Wilmot, *Innovation: The Five Disciplines for Creating What Customers Want,* page 150

202–203   "MIT's licensing office—called the Technology Licensing Office . . ." Personal interview, Dr. Robert Langer and Sarah Miller Caldicott, September 2006.

204   "A good many inventors try to develop things life-size, . . ." Dyer and Martin, *Edison, His Life and Inventions,* Chapter XXIV.

206   "seriously retarded by an inability to obtain competent engineers." Paul Israel, *Edison: A Life of Invention,* page 224.

206   ". . . accompanied by Kruesi, Bergmann, and others, . . ." and "In stores and business places throughout the lower quarters of the city . . ." Friedel and Israel, *Edison's Electric Light: Biography of an Invention,* page 222.

207   Details of how Edison donated equipment to Columbia University in 1882 to establish a school of Electrical Engineering were provided by Dr. Robert Rosenberg, former head of the Edison Papers Project at Rutgers University, in a personal interview with Sarah Miller Caldicott, March 2007. MIT is believed to be the first university to offer a degree in Electrical Engineering, followed by Cornell and Columbia.

207   "There are two types of creativity . . ." As quoted in *Let My People Go Surfing: The Education of a Reluctant Businessman,* Yvon Chouinard, Penguin Books, New York, 2005, page 97.

207   "moving from one to a thousand requires a long journey . . ." Carlson and Wilmot, *Innovation: The Five Disciplines for Creating What Customers Want,* page 201.

208   "In the 1950s, my husband and lab director, Dr. Alfred Free, devised . . ." Personal interview, Dr. Helen Free and Sarah Miller Caldicott, August 2006.

209   "Spectrum Brands uses what's called a 'stage-gate' new product development process . . ." Letter from Tom Quick to Sarah Miller Caldicott, December 2006.

210   "If you go back to the 1972–73 time frame . . ." Personal interview, Dr. Robert Kahn and Sarah Miller Caldicott, November 2006.

211   ". . . Edison name had come to serve as a form of assurance . . ." Paul Israel, *Edison: A Life of Invention,* page 301.

212   "No matter how good a machine should be invented by another . . ." Paul Israel, *Edison: A Life of Invention,* page 301.

212   "Though the fifteen-year old publisher had his problems with grammar and spelling, . . ." Lawrence Frost, *The Thomas A. Edison Album,* page 31.

212   "Edison is the Aladdin's lamp of the newspaper man. . . ." Paul Israel, *Edison: A Life of Invention,* page 372.

213 "... homespun humor and forceful statements ..." Paul Israel, *Edison: A Life of Invention*, page 354.

213 "He is fond of startling his hearers with extraordinary statements ..." Paul Israel, *Edison: A Life of Invention*, page 372.

217 "All of the great innovators who have become market-moving brands ..." Letter from Professor Jim Clawson to Michael J. Gelb, March 2007.

220–222 The PROPAR acronym was coined by Michael J. Gelb in 1990. Research background for the PROPAR principles is available in Peter Russell's *The Brain Book*, Plume, New York, 1984

222 "If you are the champion of a new innovation, . . ." Carlson and Wilmot, *Innovation: The Five Disciplines for Creating What Customers Want*, page 129.

CHAPTER 8—EDISON'S LEGACY IN THE TWENTY-FIRST CENTURY

228 "If America were a company, freedom and exploration . . ." F. Duane Ackerman, "Innovate America," "The National Innovation Initiative Summit and Report," Council on Competitiveness, Washington, D.C., 2005, page 17.

228 ... is "'somehow losing its edge . . ." Sam Palmisano, "Innovate America," "The National Innovation Initiative Summit and Report," Council on Competitiveness, Washington, D.C., 2005, page 18.

228 "I don't think there are very many companies today that are . . ." from an interview with Gary Hamel, "Q&A with Business Visionary Gary Hamel: Make the Future Match What You Imagine," *Executive Thought Leadership Quarterly*, October 4, 2000.

CHAPTER 9—THE EDISON INNOVATION LITERACY BLUEPRINT

260 "The value of an idea lies in the using of it." Thomas A Edison, *Edison & Ford Quote Book*, page 21.

Resource:

Michael J. Gelb: www.michaelgelb.com

Sarah Miller Caldicott: www.sarahcaldicott.com

INDEX

Note: Page numbers in *italics* refer to illustrations, charts and graphs.

ABOUT THE AUTHORS

Michael J. Gelb is the world's leading authority on the application of genius thinking to personal and organizational development. He is a pioneer in the fields of creative thinking, accelerated learning, and innovative leadership. He has written ten previous books, including the international bestseller *How to Think Like Leonardo da Vinci*. His clients have included DuPont, General Electric, Merck, Microsoft, and Nike, among others. He lives in Santa Fe.

Sarah Miller Caldicott is a great-grandniece of Thomas Edison, and holds an M.B.A. from the Amos Tuck School of Business at Dartmouth. She is a twenty-year marketing veteran who has spearheaded domestic and international innovations during her tenure with Pepsico, Bayer AG, and Unilever. Sarah now heads her own innovation and marketing consultancy, which has served DHL Global Mail, Cox Enterprises, and Lucent, as well as numerous entrepreneurial firms. She lives in Chicago.